Internet and Emotions

Nothing seems more far removed from the visceral, bodily experience of emotions than the cold, rational technology of the Internet. But as this collection shows, the internet and emotions intersect in interesting and surprising ways. *Internet and Emotions* is the fruit of an interdisciplinary collaboration of scholars from the sociology of emotions and communication and media studies. It features theoretical and empirical chapters from international researchers who investigate a wide range of issues concerning the sociology of emotions in the context of new media. The book fills a substantial gap in the social research of digital technology, and examines whether the internet invokes emotional states differently from other media and unmediated situations, how emotions are mobilized and internalized into online practices, and how the social definitions of emotions are changing with the emergence of the internet. It explores a wide range of behaviors and emotions from love to mourning, anger, resentment and sadness. What happens to our emotional life in a mediated, disembodied environment, without the bodily element of physical co-presence to set off emotional exchanges? Are there qualitatively new kinds of emotional exchanges taking place on the internet? These are only some of the questions explored in the chapters of this book, with quite surprising answers.

Tova Benski is Senior Lecturer at the School of Behavioral Sciences, the College of Management.

Eran Fisher is Assistant Professor at the Department of Sociology, Political Science, and Communication at the Open University of Israel.

Routledge Studies in Science, Technology and Society

1. **Science and the Media**
   Alternative Routes in Scientific Communication
   Massimiano Bucchi

2. **Animals, Disease and Human Society**
   Human-Animal Relations and the Rise of Veterinary Medicine
   Joanna Swabe

3. **Transnational Environmental Policy**
   The Ozone Layer
   Reiner Grundmann

4. **Biology and Political Science**
   Robert H. Blank and Samuel M. Hines, Jr.

5. **Technoculture and Critical Theory**
   In the Service of the Machine?
   Simon Cooper

6. **Biomedicine as Culture**
   Instrumental Practices, Technoscientific Knowledge, and New Modes of Life
   Edited by Regula Valérie Burri and Joseph Dumit

7. **Journalism, Science and Society**
   Science Communication between News and Public Relations
   Edited by Martin W. Bauer and Massimiano Bucchi

8. **Science Images and Popular Images of Science**
   Edited by Bernd Hüppauf and Peter Weingart

9. **Wind Power and Power Politics**
   International Perspectives
   Edited by Peter A. Strachan, David Lal and David Toke

10. **Global Public Health Vigilance**
    Creating a World on Alert
    Lorna Weir and Eric Mykhalovskiy

11. **Rethinking Disability**
    Bodies, Senses, and Things
    Michael Schillmeier

12. **Biometrics Bodies, Technologies, Biopolitics**
    Joseph Pugliese

13. **Wired and Mobilizing**
    Social Movements, New Technology, and Electoral Politics
    Victoria Carty

14. **The Politics of Bioethics**
    Alan Petersen

15. **The Culture of Science**
    How the Public Relates to Science Across the Globe
    Edited by Martin W. Bauer, Rajesh Shukla and Nick Allum

16 **Internet and Surveillance**
The Challenges of Web 2.0 and Social Media
Edited by Christian Fuchs, Kees Boersma, Anders Albrechtslund and Marisol Sandoval

17 **The Good Life in a Technological Age**
Edited by Philip Brey, Adam Briggle and Edward Spence

18 **The Social Life of Nanotechnology**
Edited by Barbara Herr Harthorn and John W. Mohr

19 **Video Surveillance and Social Control in a Comparative Perspective**
Edited by Fredrika Björklund and Ola Svenonius

20 **The Digital Evolution of an American Identity**
C. Waite

21 **Nuclear Disaster at Fukushima Daiichi**
Social, Political and Environmental Issues
Edited by Richard Hindmarsh

22 **Internet and Emotions**
Edited by Tova Benski and Eran Fisher

Internet and Emotions

Edited by Tova Benski and Eran Fisher

Routledge
Taylor & Francis Group
NEW YORK   LONDON

First published 2014
by Routledge
711 Third Avenue, New York, NY 10017

Simultaneously published in the UK
by Routledge
2 Park Square, Milton Park, Abingdon, Oxon OX14 4RN

*Routledge is an imprint of the Taylor & Francis Group,
an informa business*

© 2014 Taylor & Francis

The right of Tova Benski and Eran Fisher to be identified as the authors of the editorial material, and of the authors for their individual chapters, has been asserted in accordance with sections 77 and 78 of the Copyright, Designs and Patents Act 1988.

All rights reserved. No part of this book may be reprinted or reproduced or utilised in any form or by any electronic, mechanical, or other means, now known or hereafter invented, including photocopying and recording, or in any information storage or retrieval system, without permission in writing from the publishers.

Trademark Notice: Product or corporate names may be trademarks or registered trademarks, and are used only for identification and explanation without intent to infringe.

Library of Congress Cataloging-in-Publication Data
Internet and emotions / edited by Tova Benski and Eran Fisher.
   pages cm. — (Routledge studies in science, technology and society ; 22)
   Includes bibliographical references and index.
   1. Internet—Psychological aspects. 2. Internet—Social aspects.
   3. Online social networks—Psychological aspects. 4. Emotions.
   5. Social media—Psychological aspects. I. Bensky,
Tova. II. Fisher, Eran.
   HM851.I5695985 2013
   302.23'1—dc23
   2013019225

ISBN13: 978-0-415-81944-2 (hbk)
ISBN13: 978-0-203-42740-8 (ebk)

Typeset in Sabon
by IBT Global.

Printed and bound in the United States of America by IBT Global.

To my family.
T.B.

To Roya.
E.F.

Contents

List of Figures   xi
Acknowledgement   xiii

Introduction: Investigating Emotions and the Internet   1
TOVA BENSKI AND ERAN FISHER

PART I
Theoretical and Methodological Considerations

1   Power, Identity, and Feelings in Digital Late Modernity: The Rationality of Reflexive Emotion Displays Online   17
JAKOB SVENSSON

2   Feeling Through Presence: Toward a Theory of Interaction Rituals and Parasociality in Online Social Worlds   33
DAVID BOYNS AND DANIELE LOPRIENO

3   Measuring Emotions in Individuals and Internet Communities   48
DENNIS KÜSTER AND ARVID KAPPAS

PART II
Emotions Display

4   Grief 2.0: Exploring Virtual Cemeteries   65
NINA R. JAKOBY AND SIMONE REISER

5   Islamic Emoticons: Pious Sociability and Community Building in Online Muslim Communities   80
ANDREA L. STANTON

6 Emotional Socialization on a Swedish Internet Dating Site: The Search and Hope for Happiness 99
HENRIK FÜRST

7 Emotion to Action?: Deconstructing the Ontological Politics of the "Like" Button 113
TAMARA PEYTON

PART III
Mediating Interpersonal Intimacy

8 Transconnective Space, Emotions, and Skype: The Transnational Emotional Practices of Mixed International Couples in the Republic of Ireland 131
REBECCA CHIYOKO KING-O'RIAIN

9 Send Me a Message and I'll Call You Back: The Late Modern Webbing of Everyday Love Life 144
NATÀLIA CANTÓ-MILÀ, FRANCESC NÚÑEZ, AND SWEN SEEBACH

PART IV
Mediating Emotional Space

10 Emerging Resentment in Social Media: Job Insecurity and Plots of Emotions in the New Virtual Environments 161
ELISABETTA RISI

11 Cosmopolitan Empathy and User-Generated Disaster Appeal Videos on YouTube 178
MERVI PANTTI AND MINTTU TIKKA

12 Anger, Pain, Shame, and Cyber-Voyeurism: Emotions Around E-Tragic Events 193
ALESSANDRA MICALLIZZI

13 Emotional Investments: Australian Feminist Blogging and Affective Networks 211
FRANCES SHAW

Contributors 225
Author Index 231
Subject index 233

Figures

2.1	Dynamics of interaction rituals.	35
2.2	Dynamics of parasocial interaction rituals.	38
3.1	Example of a dyad (here a male and a female) discussing the topic of relationship "breakup" in an online chat in the laboratory.	52
3.2	Another dyad (here two females) discussing the topic of relationship "breakup."	52
5.1	The standard Muslim greeting "Salaam w `aleikum" or "peace be upon you."	86
5.2	Arabic-language Islamic expression emoticons also include formulaic but personal expressions like "Astaghfur Allah" or "I ask God's forgiveness."	86
5.3	"Hijab Icon".	89
5.4 and 5.5	5.4 (left) shows a smiley saying "Ma sha' Allah", "As God has willed". 5.5 (right) shows a smiley saying "Ameen", which is close to the Christian use of "Amen".	89
7.1	Riot Games personality Nikasaur encourages viewers to 'like' their video feed.	121
7.2	Official promotional video for Riot Games' League of Legends.	122
12.1	Network analysis of the first moment of video participation (dark circled: people of Avetrana; light circled: the contester).	200
12.2	The second step of video participation called "investigation" (in dark circle: the main hubs of conversations).	201
12.3	The moment of memory in video participation.	203
12.4	Explicit referents of comments.	203
12.5	Emotion frequencies coming out quantitative content analysis (%).	204

Acknowledgement

The idea of this book emerged out of discussions in recent conferences and the interest of members of Research Network 11 on the Sociology of Emotions of the European Sociological Association. We are grateful to the Research Committee of the School of Behavioral Sciences at the College of Management—Academic Studies for their financial support in preparing the index for this book.

Introduction
Investigating Emotions and the Internet
Tova Benski and Eran Fisher

What is the nature of emotions invoked on the Internet? Can we love online? Can we mourn? What does it actually mean to "like" on Facebook? And how do emotions come into play in a Skype conversation? The book *Internet and Emotions* wishes to answer these questions and more by bringing to the fore and demarcating an emerging field of interest at the intersection of two major developments taking place in the last couple of decades. The first is the penetration of the Internet into virtually every sphere of social life. From friendship and dating to shopping and mourning, no aspect of economic, political, cultural, and social life has remained unaffected by the Internet. The Internet is no longer merely another tool that people use, but an environment within which they operate and live (Poster 1997). The second development has been the advent of the sociological study of emotions. The sociology of emotions looks back on more than 30 years of history and has experienced exponential growth as a research field in recent years. Theoretical and empirical scholarship has refined our understanding of emotions and the emotional world, and has demonstrated that emotions are present in every aspect of social life and play an important role in central processes and structures. The question of media and emotions is an enduring one, but it has become ever more acute with the Internet, since unlike previous media, the Internet allows for more elaborate modes of sharing, communicating, performance, and display—all are key ingredients of emotions.

The book collects 13 chapters that explore the emotional facets of and processes occurring within the new socio-technical environment of digital, network technology. As new emotional practices emerge, new emotional feedback loops are built, and new emotional language and manifestations crystallize on the Internet, the need also arises to study them empirically and theorize them. Questions arise concerning feeling rules and display rules in the virtual world, the role of computer-mediated-communication in the daily lives of couples, the ways in which anonymity and disembodiment affect the disclosure of emotions on the Internet, the differences and similarities among various genres of computer-mediated-communication in the affordance of emotions expression, and so forth.

DISEMBODIMENT AND SOCIAL ACTION ONLINE

One of the striking conundrums raised by considering mediated emotions is: what is the status of emotions in an environment of virtual presence and disembodiment? That is, because emotions are inextricably bound with embodiment in at least two ways. First, emotions are felt and "run" through the body. Indeed almost all sociological definitions of emotions include a psycho-physical bodily component that is usually addressed as 'feeling' (Ben-Ze'ev 2001; Fontaine et al. 2002; Lazarus and Lazarus 1994). Arlie Hochschild (1983, Appendix A) claims that emotions are one of the most important biologically-given senses. Like all other senses, it is the means through which we learn about our relationship to the world, and therefore it is essential for survival. It connects external and internal reality by sending signals that are interpreted by the self, and alerts people to prepare for action, through the body. Second, emotions are perceived as reactions to external events, particularly actions by others, which affect the person and mostly involve social interactions in embodied face-to-face situations. Barbalet (2002), for example, conceives of emotions as an experience of involvement in an event, a situation, or a specific person. It is the immediate (i.e., un*media*ted) contact of the 'self' with the world through involvement, which very often involves social interactions. Most sociological theories of emotions assume the embodied presence of two or more partners in the interaction event (Kemper 1990) or interaction ritual chains (Collins 2004). In sum, emotions are perceived as interactional and embodied.

But emotions are key not only to interactions at the micro-level of individuals, but also to the shaping and reshaping of macro social structures such as cultural norms, political structures, and economic processes. And here, too, the question of an embodied co-presence is central. Randall Collins (2004), relying on Durkheim's description of the ways solidarity is produced in society (Durkheim 1912/1995), presents a theory of Interaction Ritual (IR) chains. An IR is constructed by assembling human bodies in face-to-face interaction. Emotions evolve throughout the ritual and as a result of the ritual. As a result of a successful IR, solidarity can be reaffirmed and reproduced around traditional norms and symbols, but it can also lead to the construction of new symbols and new norms, hence leading to social change. Why is embodied co-presence so vital for a successful IR to take place? Because on top of allowing the communication of abstract information, face-to-face interactions also involve additional bodily ingredients, such as a rapid back-and-forth of micro-behaviors (e.g. voice tones and rhythms, bodily movements).

The question arises, then: what happens to emotions, a key ingredient of interactions and social solidarity, in our technological age, where so much interaction is mediated by new media forms, such as text messages, Facebook posts, and Skype conversations? What happens in a mediated, disembodied environment, without the bodily element to set off a process of building IRs, thus contributing to social solidarity and social change? Are there new kinds of IRs with new forms of solidarity taking place on the Internet? Or does online interaction completely lack the kind of emotional component which would allow for successful IRs to take place? Empirical evidence thus far points to a mixed answer: the Internet does allow for IRs to take place, but with weaker effects: "collective effervescence never rises to very high levels; and solidarity, commitment to symbolism, and other consequences continue to exist but at a weakened level" (Collins 2011).

As the chapters in this volume suggest, the question of presence and embodiment online is complex. Taken together, the chapters suggest that rather than talking about 'disembodiment' it is more useful to think of new modes of embodiment and co-presence on the web. For example, Rebecca Chiyoko King-O'Riain suggests the term "transconnective space" (Chapter 8, this volume), and suggests that the media richness (Daft and Lengel 1984) provided, for example, by Skype can itself serve as a prosthetic membrane, substituting the actual body. We might, then, need to think of different and multiple types of embodiment taking place in various media forms. This point relates to the unique characteristics of the Internet. The Internet is a complex, multi-faceted technology that integrates many media forms; it is an ensemble of media. Hence, when we talk about 'the Internet' we, in fact, talk about different and varied information and communication technologies. Writing or reading a blog, querying a search engine, emailing, reading a newspaper online or reading news from multiple resources using RSS feeds, using Skype to communicate with friends or colleagues—all of these entail different media and different experiences. Moreover, these various media tend to converge online—watching 'television' or 'movies' online, speaking over the 'phone' using Skype, or reading a 'newspaper' online (Jenkins 2008). This is highly consequential to emotions. To the extent that emotions involve the ability to communicate minute micro-behaviors (such as vocal gestures or facial expressions), then a higher media richness (such as provided by Skype) may facilitate a kind of Interaction Ritual which Facebook may not (see: Collins 2011).

THE SOCIAL CONSTRUCTION OF EMOTIONS ONLINE

Another fundamental tenet of the sociology of emotions, pertinent to our concern here, is the notion that emotions are not merely subjective and individual but also social; they are not merely sensory and cognitive experiences but also involve normative and performative aspects. Emotions are filtered and constructed through social norms which dictate how one should feel (feeling rules), and behavioral codes which dictate how one should express these emotions (display rules; Hochschild 1983, 1990). Behavior is not simply a mode of externalizing emotions; it also informs people of how they feel, and can facilitate or curb sensory experiences. For example, feeling rules and display rules of mourning in Western cultures commonly delimit the expression of grief temporally and spatially, so that it is more acceptable to grieve in cemeteries than in other places. Such rules, then, dictate the display of grief, which in turn may influence how grief is experienced and dealt with. In the context of this volume the question that emerges is: what happens when the context where behavior takes place changes? How, for example, a different set of behaviors that emerges in online cemeteries changes how emotions of grief are experienced, displayed, and operate in interpersonal and public contexts? These themes are taken by Nina Jakoby and Simone Reiser in this volume.

As a technology, the Internet can be characterized as relatively decentralized and distributed (compared with the mass media), allowing users to interact with the media, participate in the production of its content, and collaborate with others. But the Internet should not be thought of merely as a technology; rather, it is a socio-technical system comprised not only of machines, but also of human actors, regulations, social norms, and social structures. In the same way that the mass media cannot be understood outside the context of mass society, mechanical technology, the big factory, and the political system within which it operates, so it is appropriate to situate the Internet within its social context. The overarching concept which seems both to account for the multiple characteristics of the Internet and relate these to the social context is the notion of networks; indeed, the Internet can be seen as an expression and an enabler of network society.

The development and deep penetration of the Internet has occurred concurrent with deep social transformations in the last few decades, resulting in the emergence of the network society (Castells 1996). The network society is characterized by increased connectedness and interdependence between nodes. This is highly consequential to society, economy, politics, and culture. It also has two important implications for the study of the Internet and emotions. The sociology of emotions asks us to look at emotions as taking place *between individuals* and within *social contexts*. Networks entail increasing connectedness between individual nodes (for example, through emails or Skype conversations), as well as new social spaces where emotions are invoked and performed (exemplified in public and semi-public spaces such as blogs, forums, and social networking sites).

Social life itself has become more fragmented, distributed among different places, experiences, and friends. In such a liquid environment, as Bauman (2000) puts it, online communication may help overcome the difficulties of sharing emotions face-to-face. This is apparent, for example, in the case of work life. As Elisabetta Risi suggests in this volume, "experiences of [job] insecurity are 'dispersed,' since precarious workers are themselves scattered in different work sectors and companies" (Risi, this volume). The Internet can serve as a locus for displaying and experiencing emotions in a realm of employment which has become more networked, where workers are more atomized and dispersed. This, in turn, might uncover and highlight the societal origins of these emotions, and allow for political action to take place. In other words, the Internet is both an expression of atomization and a means to overcome it. A similar idea is found in Henrik Fürst's analysis of a Swedish dating site (Fürst, this volume), which is seen by users as one of the possible social arenas for meeting future partners and love. Users are drawn to the dating world looking for the alleviation of feelings of loneliness and in order to gain a feeling of self-worth through the evaluation of others.

The Internet, with its multiple technologies, also gives rise to a variety of different types of virtual spaces and practices. Some of the practices like mobile phones, SMS, or spaces created by Skype conversations retain the characteristics of the private sphere, where public access is restricted, giving rise to one-to-one channels of communication. Other types of Internet media are more public in their nature, such as community forums, blogs, vlogs, and international appeal sites, creating one-to-many channels of communication, often communicating with individuals unknown and unfamiliar to the author. As a result, some of these spaces defy the private/public binary: on the one hand some of these sites are open to all, making it public, but on the other hand these sites might feature very personal information and emotion displays, thus turning these spaces into the most publicly-created private spaces we can encounter.

Notwithstanding the popular image of the spaces that emerge through various digital communication technologies as affording spontaneity, wide public accessibility, freedom of expression, and horizontalism, some of the chapters in this volume show otherwise. We will bring here only three examples that make the point in clear voice. First, Fürst, in his study of a Swedish dating site shows that perhaps the most clearly emotion-directed sites on the Internet are in effect well-structured, highly-controlled spaces. His study reveals the phantasmagoria of happiness—or rather of a hope for happiness—built into the site, commercialized, and regulating commitment of users. Second, Jakob Svensson, in his chapter on the rationality of emotion displays online, focuses on the connection between emotions management online and the new constellations of power relations in the network society. Contrary to the image of the web as a horizontal, flat space which allows equal access to all participants, Svensson highlights how power relations are reproduced on the web. Emotions display online require elaborate forms of reflexivity and identity negotiation. The middle and upper classes, he argues, have greater access to acquiring these skills. They develop what might be termed after Bourdieu (1984) *digital* social capital which can be translated into other forms of capital. Finally, third, contrary to the common view of open accessibility, many online spaces, from talkbacks to forums, are carefully monitored. For example, Frances Shaw (this volume), in her analysis of feminist blogs in Australia discusses the efforts invested in the moderation of the site to create a "safe space" where feminist counterhegemonic ideas can develop freely, and where strong emotions, not easily accepted by codes of politeness and civility in the mainstream public sphere, can be freely expressed and shared. But this 'freedom' and 'openness' are achieved through restricting access to the forum.

The Internet offers a unique place to study emotions, not only for empirical and theoretical reasons, but for methodological reasons as well. It can be thought of as a unique laboratory for the study of emotions for two key reasons. First, the Internet is a fertile ground for a huge diversity and amount of communication of all sorts and from a large and diverse group of people. Much of that communication is emotional, reflecting immediate feelings, sometimes, as they occur—most use of social media such as Facebook and Twitter is now occurring on mobile devices. Second, these communication acts are all registered. Data may not always be readily available because of commercial reasons or privacy concerns. But when communication data become available, it is relatively easy to analyze since it is likely to be relatively complete and includes meta-data such as time and location, and at times other pieces of important demographic information about the authors of the data, such as gender, education, or online behavior. Indeed, much Internet and emotions research takes advantage of these characteristics, analyzing online communication through either quantitative or qualitative methods. A combination of both quantitative and qualitative methods is particularly interesting, given the unique nature of data online: it is human communication, which is hard to reduce solely to quantitative features (hence the use of qualitative methods), yet it is comprised of huge amounts of speech acts (hence the use of quantitative methods). Such analysis can account for large quantities but still retain minute interpretive qualities.

THE ORGANIZATION OF THIS BOOK

The book is comprised of 13 chapters, divided into four parts. Part I of the book features three chapters which attend to theoretical and methodological considerations. Chapter 1, Power, Identity, and Feelings in Digital Late Modernity, by Jakob Svensson, explores how emotions are managed and displayed online, and how this new environment for emotions display forms new relations of power. Svensson associates the unique character of managing emotions online with the "reflexive identity negotiation of digital late modernity." He sees the management of emotions online within the long lineage of what Norbert Elias (2000) called "the civilizing process," where emotions are taught to be contained and manipulated. Social networking sites (SNS) put further strain on individuals to manage their emotions in order to increase their social capital, and gain recognition and status from peers. As Svensson puts it, "reflexive identity negotiation, an urge to display a coherent and attractive self, guides emotion work online" (Svensson, this volume).

Svensson asks how the display of emotions online is rationalized and for what purpose. Building on Hochschild's conceptions of emotions display as a form of capital in labor (1983), Svensson shows how feelings may be managed in SNS for identity purposes: "To answer the increasingly important question in late modernity of who am I, people turn to feelings in order to locate themselves. . . . On SNS users . . . negotiate their identities and group belongings through displaying feelings" (Svensson, this volume). Unlike smiles for flight attendants, feelings on SNS rarely have any monetary exchange-value for users, but they can be exchanged for recognition and status. The network architecture further intensifies the importance of connections and connectivity between people: "The rise of (semi)public management of feelings can thus be understood in light of an increasing need to account for others and co-exist in a society where connections between people multiply and connections in themselves become more and more important" (Svensson, this volume).

In Chapter 2, Feeling Through Presence, David Boyns and Daniele Loprieno employ categories from Interaction Rituals Theory to the study of emotions online. They argue that the Internet is a particularly vibrant setting "in which emotional interactions are prominent in social interaction" (Boyns and Loprieno, this volume). As aforementioned, Collins' theory of Interaction Rituals posits emotions as central to social cohesion and social change, but predicts that the more technologically-mediated our interactions become, the less likely they are to invoke the same levels of emotional energy as direct face-to-face interactions, for lack of physical co-presence. The chapter tackles the central issue of virtual presence by asking to what extent technologically-mediated interactions can transcend the lack of bodily presence.

The chapter argues that a form of disembodied, parasocial presence can in fact take place, but it is conditioned on two "illusions" enabled by digital technology: an illusion of intimacy, and an illusion of non-mediation. "Through experiences of 'parasocial presence,'" the chapter argues, "individuals become emotionally immersed in technologically mediated interactions, their interactions develop qualities that simulate the experience of co-presence and take on the more general characteristics of IRs" (Boyns and Loprieno, this volume). The chapter provides empirical evidence to the power of the Internet to offer emotionally-charged experiences. Most notably, research into online worlds (particularly games) finds that they provide immersive experiences which invoke a high level of emotional engagement and a strong sense of presence and intimacy.

Chapter 3, Measuring Emotions in Individuals and Internet Communities, by Dennis Küster and Arvid Kappas, provides an overview of fundamental methodological challenges in the study of emotions online. It raises key questions for anyone researching emotions online, such as the difficulty of a direct, simple "reading" of emotions online (e.g., a smiling emoticon might actually indicate feelings of embarrassment), and explore possible solutions. The Internet offers a unique opportunity to collect and measure emotions' expressions on it. It is a unique social setting in that records are

8 *Tova Benski and Eran Fisher*

kept and create a rich database. Studying emotions on the Internet "offers vast amounts of data and greater ecological validity than most laboratory experiments" (Küster and Kappas, this volume). The chapter outlines three areas of emotions measurement, each requiring its own unique methods, and each revealing a different facet of the intersection of the Internet and emotions. First, we can investigate large amounts of emotional content readily available online (through qualitative or quantitative content and data analysis). Second, we can inquire into the subjective emotional experience of users (using self-reporting, through interviews or questionnaires). And third, we can record bodily responses indicating emotional states in real-time Internet use.

Part II of the book focuses on emotions display online. It features four chapters, each dealing with new ways by which emotions are displayed in a mediated environment. Chapter 4, Grief 2.0, by Nina Jakoby and Simone Reiser, investigates the new emotion rules of grief which emerge online. The chapter employs virtual cemeteries as a "methodology lab" to explore, test, and challenge assumptions of symbolic interaction theory regarding grief, particularly its insistence on physical parameters. The authors suggest that virtual cemeteries set mourners free from feeling rules in four ways: they are unrestricted by time limitations, and they allow spiritualization and afterlife beliefs, mourning for untraditional "family" members (such as a secret lover), and direct conversations with the dead. At the same time, while virtual cemeteries offer a social space that allows more freedom from conventional feeling rules of grief, they also enforce their own feeling rules, such as an infinite process of mourning. The chapter highlights how the unique spatiality and temporality that exists on the Internet transforms grief, as much of the feeling rules and practices involved in mourning and grief are tied with elements of space and time, indicating, for example, where grieving is acceptable and to what duration after the death of a loved one. The chapter concludes that the Internet has not given rise to a new form of grief, but it has influenced how grief is expressed: "The feeling rules exist prior to the medium by which emotions are expressed . . . The Internet only influences the way emotions are expressed but not the experience of the emotion itself" (Jakoby and Reiser, this volume).

The problem of displaying emotions online and the means to overcome this problem are also taken up in Chapter 5, Islamic Emoticons, by Andrea Stanton. By carefully interpreting the use of emoticons within a specific online social setting, the chapter shows how emoticons "help ameliorate the disconnection caused by the anonymity and disembodiment of online avatars" (Stanton, this volume). By thus, it is also contributing to a central conundrum of the Internet and emotions: how can emotions be displayed without bodies (physical and identifiable)? The chapter explores emotional display in online Muslim communities, giving us an intimate and detailed account of how mediated emotions operate. It studies the acceptance and use of specialized "Islamic" emoticons in online Muslim communities by both clerical authorities and regular users. The chapter finds that Islamic emoticons operate not simply as communicative indicators of emotions but that they also convey other messages, central to the creation and maintenance of a religious public sphere. "Islamic emoticons," Stanton argues, "help provide their users with religious credibility. They offer a distinctive, supra-textual demonstration of users' commitment to a Muslim identity and personal piety" (Stanton, this volume).

The centrality of emotions display in social situations is perhaps best epitomized in romantic relations, particularly in the initial stages of courting. Chapter 6, Emotional Socialization on a Swedish Internet Dating Site, by Henrik Fürst, sets this question against the backdrop of dating sites, dedicated to instigate the initial "chemical reaction" between future couples. It shows how the specific characteristics of this online environment—both highly rationalized and overflowing with emotions—encourage a specific emotional state. Fürst casts Internet dating sites as arenas that control and shape the emotional ordering of users, regulating, in other words, users' spontaneous emotions. Users of dating sites go through a process of emotional socialization concerning hope and happiness, which, in turn, reproduces user's engagement with the site and commitment to it. Dating sites, then, are not neutral arenas where users interact and communicate; rather, they are well-structured and controlled emotions' spaces. Internet dating companies control what Fürst calls the "imaginary life" of service users. At the center of this structuring is the feeling of hope. Internet dating sites make every attempt to keep daters "in a state of hope through stabilizing the emotional ordering, as hope is to be constantly present and reproduce the (economically) valuable Internet dating activities" (Fürst, this volume). Dating site companies regulate feelings of hope for happiness to keep their customers committed.

The last chapter in Part II questions the very ability of the Internet to allow a genuine display of emotions. Chapter 7, Emotion to Action?, by Tamara Peyton, attends to one of the most iconic expressions of emotions display in contemporary web culture: the "like" button of Facebook. Exploring empirical examples of "liking," and setting this mundane action within the context of contemporary capitalism, the chapter argues that the very discursive ontology of emotions has been transformed in the context of the Internet. The meaning of the term "like" has "moved away from the realm of the emotive internal life of individuals and into the realm of the discursive public sphere of societies" (Peyton, this volume). Rather than an indication and an expression of internal emotions, "liking" has become an action entangled within the web of communicative capitalism. Not only does "liking" have varied meanings, but these are in a constant state of flux. To "like" something means different—and sometimes conflicting— things for different people and in different contexts. Through the analysis of the meaning of the "like" button, the chapter shows a process of versatilization of emotions' metaphors, also involving a radical change in the field of action. What many of these different meanings have in common is that pressing the "like" button is no longer an expression of an internal emotion. Instead, it connotes connections between nodes in the networked society. Perhaps this is the way that nodes connect? If so, this, in effect, calls into question the action/emotion dichotomy, e.g., "liking" (i.e., clicking the "like" button on Facebook) can precede the emotion, or be completely independent of it.

Part III of the book, Mediating Interpersonal Intimacy, attends to the role of the Internet as a means of interpersonal communication. It features two chapters that investigate how emotions are mediated in one-to-one, private communication settings. Chapter 8, Transconnective Space, Emotions, and Skype, by Rebecca Chiyoko King-O'Riain, investigates how Skype allows global families to communicate and share in meaningful and intimate ways, notwithstanding physical distance. At the same time it underscores the frustration of users with emotional and bodily limitations posed by Skype. Hence, the chapter looks at the practice of 'doing' relations online. Rather than being merely a media for communication, Skype is understood to create a "transconnective space, which attempts to create a new 'semi-virtual' space . . . by using Skype to create ongoing co-presence, by streaming emotions rather than storing them" (King-O'Riain, this volume). However, the chapter suggests that the richness of Skype—its ability to deliver both audio and video information—is a double-edged sword for emotional expression. It allows a broader channel for emotions to be expressed, and facilitates greater intimacy. But such intimacy and emotional charge might prove too overwhelming for users to bear when they communicate with a loved one who is, in effect, not present and might even be absent for a long period of time.

Chapter 9, Send Me a Message and I'll Call You Back, by Natàlia Cantó-Milà, Frances Núñez, and Swen Seebach, investigates the webbing of communication technologies in the everyday love life of couples. It asks how digital communication practices change everyday intimacy and romantic relations of couples. Based on interviews with Spanish and German couples, the chapter questions how the "love habits" of couples—"the shape and content of the relationships that bind couples together"—are transformed by new communication tools. It argues that by opening up new "possibilities to remain in touch despite the distance," digital communication technologies are "webbing new forms of living and experiencing partnership, and subsequently enabling new possibilities of defining, experiencing, and expressing commitment" (Cantó-Milà et al., this volume).

The chapter goes beyond the preliminary dating phase of relationships, which received much attention in academic research and popular discourse, and focuses instead on couples already in a relationship. It examines their use of digital communication technology and its effects on their relationship. It finds that not only do digital media devices play a role in most contemporary love relationships, but that they change certain forms of emotions; on the

Internet and on mobile devices, emotions are expressed differently. Thus, for example, mobile devices accelerate the tempo of communication, and different applications and devices allow for different emotional modalities. Especially in relations that are heavily dependent on digital communication, the chapter suggests, the medium becomes central in shaping the emotional universe of the relationship. Ultimately, the chapter concludes that notwithstanding its heavy use and immense advantages for relationships, electronic communication does not, in fact, allow "the same empathy and immediacy as fact-to-face communication" (Cantó-Milà et al., this volume).

Whereas Part III focuses on the emotional channel that the Internet facilitates between individual users, the next and last part of the book broadens the gaze and looks at the Internet as a mediated space which allows many forms of communal, semi-public, and public communication. Part IV, Mediating Emotional Space, features four chapters that ask about the emotional public space that emerges online. Chapter 10, Emerging Resentment in Social Media, by Elisabetta Risi, investigates the communication dynamics in online forums of Italian workers afflicted by job insecurity. The chapter shows the dual tendencies of the Internet. On the one hand it facilitates social communication, allowing precarious workers to express their resentment toward their working life situation. On the other hand, such communication becomes an end in itself, inhibiting a more political action, oriented toward changing this social malaise. The sharing of negative emotions, such as resentment, can become "a way to access a socially constructed reality and its effects can therefore go beyond the individual level" (Risi, this volume), leading in fact to a political action. The chapter points to the potential of the online communication of emotions to lead to political and social action, but also acknowledges that such potentials may not materialize. "[T]he online sharing of resentment," says Risi, "seems unable to strengthen . . . weak ties beyond the mere act of joining online environments. . . . [I]n the Italian social context the emotional constellations of precarious workers have only occasionally eased the way from online proposals to social projects or actions." Hence, the public mediation of emotions of resentment due to increasing job insecurity results in a dual trend: on the one hand, in these communities, "the energy of resentment is overexposed and amplified," on the other hand, this form of public 'venting' also dissipates these emotions, thus hampering the political potency of these emotions.

A more optimistic view—especially for scholars of social movements and social change—is suggested in Chapter 11, Cosmopolitan Empathy and User-Generated Disaster Appeal Videos on YouTube, by Mervi Pantti and Minttu Tikka. The chapter investigates the extent to which the Internet facilitates the mobilization of emotions for political goals. It shows how "cosmopolitan empathy" is generated in post-humanitarian society through the participation of ordinary people in disaster communications. The chapter analyzes video blogs (vlogs) and video montages generated by ordinary people and posted on YouTube, appealing to audiences in the West to get involved in the earthquake and tsunami in Japan and East Africa's famine, respectively. Because of the "direct address, the everyday language, the private *mise en scene* and the amateurishness of the production" (Pantti and Tikka, this volume), vlogs offer a unique social space for the communication of emotions, facilitating an intimate appeal with a mass audience reach. The chapter explores the possibilities of engaging a global media audience to create a cosmopolitan emotional empathy, thus "extending the boundaries of collective care and moral community" (Pantti and Tikka, this volume). Being a global space, new media "challenge the nation as the ultimate moral community and its monopoly over public emotions," with "ordinary people . . . now increasingly appearing as independent agents of humanitarian communication and functioning as 'educators' for the emotional and moral response" (Pantti and Tikka, this volume). Ironically, however, the chapter argues that these YouTube videos create an emotional circle which in fact excludes the victims of the disaster, affirming Western audience's "own humanity, rather than others' suffering" (Pantti and Tikka, this volume).

Chapter 12, Anger, Pain, Shame, and Cyber-Voyeurism, by Alessandra Micallizzi, further delves into the characteristics of the Internet as a new emotional public space. It explores how the Internet can serve as a space for the social sharing of emotions. The chapter analyzes the online discussions following the disappearance of a young Italian girl. Micallizzi offers a detailed analysis of the discussions, using both quantitative and qualitative narrative analyses to identify the emotional characteristics of the discourse. The chapter argues that the online event best be understood as a case of cyber-voyeurism. Such an event revolves on the "social sharing of emotions," which involves "others in the processing of personal emotive experiences, especially the negative ones" (Micallizzi, this volume). The chapter concludes that "virtual environments . . . are comfortable contexts where people can cope with trauma and negative emotions by sharing with other people" (Micallizzi, this volume). Through the use of different techniques of self-expression and representation, characteristic of the net, users can create a space that is "privately public or publicly private," where they "can experience emotive projection and self-disclosure with a therapeutic effect" (Micallizzi, this volume).

Lastly, in Chapter 13, Emotional Investments, Frances Shaw highlights the political affordances of emotions shared on the semi-public sphere of feminist blogs. According to the chapter, anger, frustration, outrage, disgust, and disdain are conceived not merely as individual emotions to be curbed and soothed, but as affective indicators, instigating politically-motivated acts, such as expressing unwarranted and unpopular views. Feminist bloggers employ various mechanisms to make their blogs an emotionally-protective space, such as careful moderation of responses, exclusion of trolls, censorship of expressions of hate, and a "trigger warning" which alerts women readers of content that may be emotionally-charged. The incessant maintenance of these mechanisms also renders blogs a support network for their participants, a community. But, as the chapter insists, this does not dismiss the political potentialities of these blogs, "the political nature of affective community," as Shaw calls it.

The chapter sees blogs as spaces that allow feminists to turn their anger—directed at the mainstream political discourse—"into political claims to other women." "Through this process," she argues, women feminist bloggers "were able to articulate particular claims in response to mainstream discourses that made them angry. . . . These articulations provided the justifications for further confrontations with the mainstream, for not only them, but also the others that they have now armed with the discourse to do so" (Shaw, this volume). Similarly to the arguments advanced by Risi (this volume), this chapter, too, shows us that when mediated through blogs, emotions of anger and frustration of excluded or disenfranchised political subjectivities can be articulated into an effective political discourse. "Women found the articulation of their political feeling to be most powerful when other women could identify with that feeling." The chapter illuminates how emotionally-charged political discourse—which cannot be easily contained within a general public sphere—curves a safe space online, facilitating the development of a counterhegemonic discourse.

REFERENCES

Barbalet, J.M. (2002) *Emotions and Sociology*. Oxford: Blackwell.
Bauman, Z. (2000) *Liquid Modernity*. Cambridge: Polity.
Ben-Ze'ev, A. (2001) *The Subtlety of Emotions*. Israel: Zmora-Bitan.
Bourdieu, P. (1984) *Distinction: A Social Critique of the Judgment of Taste*. Cambridge, MA: Harvard University Press.
Castells, M. (1996) *The Rise of the Network Society*. Oxford: Blackwell.
Collins, R. (2004) *Interaction Ritual Chains*. Princeton, NJ: Princeton University Press.
Collins, R. (2011) Interaction Rituals and the New Technology. *The Sociological Eye, Writing by Sociologist Collins*. Retrieved December 7, 2012 from http://sociological-eye.blogspot.co.il/2011_01_01_archive.html
Daft, R.L. and Lengel, R.H. (1984) Information Richness: A New Approach to Managerial Behavior and Organizational Design. In L.L. Cummings and B.M. Staw (Eds.), *Research in Organizational Behavior*, 6, 191–233. Homewood, IL: JAI Press.
Durkheim, E. (1912/1995) *The Elementary Forms of Religious Life*. New York: The Free Press.
Elias, N. (2000) *The Civilizing Process. Sociogenetic and Psychogenetic Investigations*. Oxford: Blackwell.
Fontaine, J.R.J., Poortinga, Y.H., Setiadi, B. and Markam, S.S. (2002) Cognitive Structure of Emotion Terms in Indonesia and The Netherlands. *Cognition & Emotion*, 16(1), 61–86.
Hochschild, A. (1983) *The Managed Heart: Commercialization of Human Feeling*. Berkeley, CA: University of California Press.
Hochschild, A. R.(1990) Ideology and emotion management: A perspective and Path for future research. In T. D. Kemper (Ed.), *Research agenda in the sociology*

of emotions. SUNY series in the Sociology of Emotions (pp. 117–142). New York: SUNY.

Jenkins, H. (2008) *Convergence Culture: Where Old and New Media Collide.* New York: NYU Press.

Kemper, T. (1990) Social relations and emotions: a structural approach. In T. D. Kemper (Ed.), *Research agenda in the sociology of emotions.* SUNY series in the Sociology of Emotions (pp. 207–236). New York: SUNY.

Lazarus, R.S. and Lazarus, B.N. (1994) *Passion and Reason: Making Sense of Our Emotions.* New York: Oxford University Press.

Poster, M. (1997) Cyberdemocracy: The Internet and the Public Sphere. In D. Porter (Ed.), *Internet Culture* (pp. 201–218). New York: Routledge.

Part I
Theoretical and Methodological Considerations

1 Power, Identity, and Feelings in Digital Late Modernity
The Rationality of Reflexive Emotion Displays Online

Jakob Svensson

INTRODUCTION

As digital natives grow into adolescence, social network sites (SNS hereafter) are becoming one of the most important places for sociability in connected societies. A recent study showed that young people in Sweden (12–16 years old) spend more time with their friends online than they do offline (Medierådet 2010). Thus, digital technology in general, and SNS in particular, are becoming indispensable platforms for socializing and organizing activities of all kinds. In fact, socializing without the Internet and mobile accesses to it is probably unimaginable for many youngsters today. One outcome of this shift toward the online as *the* social scene is the displacement of emotion displays from more private and group confined settings to become performed on the screen, visible for an entire network of selected peers to watch, comment, judge, and perhaps share with others. Together with this displacement, we see an increase of the kind of feeling management and emotion control that Hochschild (1983/2003) found in her groundbreaking study of smiling flight attendants. Feelings have been understood as signalling our hopes, fears, and expectations, thus carrying with them information on who we are. Hochschild observed how this signalling function became impaired when the management of emotion was socially engineered. What she discovered among smiling flight attendants in the early-'80s is now happening on a full scale online. The purpose of this chapter is to understand the rationality of emotion displays on SNS. While the flight attendants in Hochschild's study were smiling in order to convey feelings of safety and for customers to feel positively for the chosen airline, the question in this chapter is for what purposes feelings are managed online, i.e., what is the rationality of emotion displays online?

In particular, the aim of this chapter is to link the practice of feeling management to emerging forms of power relations in digital and late modernity. We can connect power to emotions in different ways. Hochschild (1983/2003: 158), for example, argues that the act of managing feelings in itself, has been confined to upper and middle class settings to a large extent. This is changing when SNS start to claim a dominant position as the arena for sociability and emotion display among different strata of youngsters in network societies. Emotions can also be linked to power via the norms and values they convey. Elias (1939b/1998), for example, argued that feelings not only should be understood as a mirror of some kind of self, but also carrying with them information and clues about accepted and preferred behaviors of a given society in a given situation. Hence, what happens when (a) an upper/ middle class activity is starting to be performed by different strata in network societies, and (b) what can we say about the norms and values of a society in which emotions are increasingly displayed online? Through attending to these two questions the chapter attempts to address the larger issue of understanding the rationality of contemporary emotion displays.

Why is it relevant to investigate emerging forms of power relations? Accounts of an emerging network society (see Castells 2000: 519, and van Dijk 2006: 19–20, for a definition) have been accompanied by a considerable number of claims when it comes to power. It is argued that people will be able to communicate and engage directly with one another on a global scale, bypassing traditional producers and distributors of information (Bruns 2008: 14). The increasing use of the network metaphor indicates for many a shift in power relations, from top-down hierarchies to more horizontal, flexible, and flatter social organization (Benkler 2006; Bruns 2008; Shirky 2009). However there is no reason to believe that increasing practices of social organization in networks will cause a society without power (see, for example, Elias 1939b/1998). Rather, what we are witnessing is a shift from more tangible and easily observable relations of more hierarchical power relations to more non-transparent and complex horizontal relations of power (or soft power, see Bakir 2010: 7). Relationships are multiplying since the network implies an emphasis on connections and connectivity between people (Miller 2008; van Dijk 2006: 24). And if we adhere to a conception of power as processes that take place *between* people (Elias 1970/1998: 115–116; Foucault 1979/1994: 324), it becomes interesting and relevant to investigate power relations in the transition toward a network society in which relations and connections between people are brought to the fore.

This chapter will start with a background account of our digital and late modern society and the rise of online social networking. The idea is that technology and society mutually inform each other and that processes of individualization and identity are underlined both in late modern societies as well as in technological developments toward online social networking. This will be followed by a section on feelings and their management and how to understand the increase of emotion displays on SNS against the backdrop of individualization and identity in digital late modernity. In this approach of linking feeling management online with processes of individualization and identity negotiation, the discussion of power will both revolve around how emerging communication practices in network societies convey norms and values that discipline users to negotiate feelings and display emotions, as well as how different strata are unequally equipped for such practices. The chapter will end with a discussion about how these developments toward (semi)public management of feelings can be understood as part of the civilizing process, linking the discussion on SNS and feelings in digital late modernity to Elias' ideas of rationality, emotion control, and self-constraint.

BACKGROUND: THE RISE OF ONLINE SOCIAL NETWORKING IN DIGITAL LATE MODERNITY

The Internet has now been around for over two decades. What is often considered new are developments toward mobile access to the Internet (Rheingold 2002) and Web 2.0. (see Leaning 2009: 30). Web 2.0 constitutes the architecture of online user participation and encompasses social software that enables many-to-many publication (Bakir 2010: 2–3). This is often referred to as social media and social network sites (SNS). Whereas social media is largely used for describing increasing possibilities of interactivity afforded by Web 2.0 technologies (commenting functions etc., see O'Reilly 2005), SNS allow the user to articulate his/her contacts and relationships and to make them visible to other users. SNS are thus defined as web-based services allowing individuals to create a (semi)public profile, connecting this profile to other users (often self-selected), whose contacts in turn will be made accessible by the service (Ellison and boyd 2007: 2). The rise of SNS has been remarkable. In 2009 the top 20 Internet sites included six SNS (Facebook, YouTube, Myspace, Twitter, RapidShare, and QQ.com, see Bakir 2010: 155). There is no doubt that these sites and mobile accesses to them have changed the communication landscape profoundly. This leads to social developments with its own significance beyond the technical aspects of digitalization. In other words, SNS are altering the way we live and socialize, and they are shaping the way things get done, providing access to information and providing us with new practices for arranging and taking part in all sorts of activities, encounters, and social agency (Dahlgren 2009: 152).

The emergence of SNS highlights the importance of connectivity in network societies. As people increasingly operate in multiple social networks (of neighbors, colleagues, friends, etc.), individuals are to a larger extent perceived as nodes and links to others (Rheingold 2002: 57, 170). This is one reason for using the network metaphor to describe contemporary Western societies, to underline that the possibilities for connectedness have increased dramatically. This, in turn, brings practices of relationship management to the fore, which is illustrated in studies of text messaging among young people. Most of the messages sent among Japanese teenagers consisted of the intimacy-maintaining 'thinking of you' or 'whatcha doin' sorts (Rheingold 2002: 5, 16). And studies have shown that if young people do not continuously receive text messages, they feel unloved and forgotten (Rheingold 2002: 21, see also Ling and Yttri's study of Norwegian youth 2002: 149). It seems that social networking provides a kind of social glue between people engaged in it. As a consequence, Miller (2008) argues that as we move into network sociality, communication becomes 'phatic,' communication for the purpose of reassuring social bonds, to offer socio-emotional support rather than exchanging information (see also Katz and Aakhus 2002: 8). Already in 1987, Rice and Love found that socio-emotional content constituted around 30% of the messages online, of which most were positive. Networks are about relations, the content transmitted in them is of secondary importance (van Dijk 2006: 95). Therefore on SNS, the most important list is the list of friends and the point of social networking is to establish and demonstrate linkages and connections, rather than to engage in dialogic communication (Miller 2008: 393).

The increasing importance of connections and connectedness is also linked to self-expression and identity negotiation. In a study of the mobile phone, informants claimed that the phone enriched their social life, furthering opportunities for self-expression at the same time as managing and remaking relationships with friends and family (Pröitz 2007). What seems to be at stake is the position within the network (see also Livingstone's study of British teenagers online 2008). In this way individuality is both fostered and dependent on the network since we most likely would be ignored without network visibility with references to other users. Hence, it seems that negotiating individuality through connectivity and network visibility are important rationales behind contemporary practices of online social networking. Being part of social networks and putting one's network connections on display become vital aspects of self-presentation, personality, and identity negotiation (Svensson 2011).

This is where I suggest the idea of a *digital late modernity* for understanding contemporary network societies. The shift toward online social networking is happening at the same time as we are experiencing societal and cultural changes toward individualization in late modernity. Individualization refers to a lacking sense of social belonging and a growing sense of personal autonomy. The collective and the traditional have faded in importance in favor of a continuous individual identity formation project (Giddens 1991). When the possibility of managing and crafting individual identities increases, the making of oneself becomes pivotal as an end in itself to strive for. A continuous emphasis of the self as something that can be managed bears upon the individual to such a degree that the self becomes a reflexive project (Giddens 1991: 32). This underlines reflexivity as a central theme in late modernity (which is sometimes referred to as reflexive modernity, see Beck 1995; Giddens 1991). SNS have made it easier for people to communicate within larger and reflexively self-made peer networks. But technology should not be understood as an independent and determining force here; it rather enables and reinforces tendencies of reflexive individualization (Leaning 2009). The mutuality of processes of digitalization, individualization, and reflexivity are clearly illustrated on SNS where individuality and personality are impossible without network visibility and references to other nodes in the network and their supposed connotations (Castells 2001: 129–133; Donath and boyd 2004; Walther et al. 2008). In this way the emerging network society is interconnected with socio-cultural changes in late modernity.

In this section I have discussed social processes toward reflexive individualization and technological developments of digitalization as mutually reinforcing each other in digital late modernity. In the next section I will address practices of emotion control and feeling management and link these to the rise of online social networking. Self-expression, identity negotiation, and maintenance are obviously connected to display and control of emotions and the self, practices that are increasingly affordable, accessible, and visible on SNS.

NORMS AND VALUES OF FEELING MANAGEMENT AND EMOTION CONTROL IN DIGITAL LATE MODERNITY

Models for understanding emotions range from essentialist definitions of emotions as biological processes, to constructivist definitions of emotions as conditioned by culture and social structures (Turner and Stets 2005). What Hochschild adds to the discussion is that feelings also can be managed and used for a purpose, something Turner and Stets (2005: 23–24) labels as part of a dramaturgical tradition of understanding emotions, i.e., emotion displays as strategic performances in front of an audience of selected others. By studying smiling flight attendants Hochschild (1983/2003) showed how feelings were used as a form of capital in labor. According to her, emotional labor is the management of feelings to create a publicly observable facial and bodily display, that in the case of flight attendants is sold for a wage and hence has an exchange value (Hochschild 1983/2003: 7). In this section I will apply her theorizing to emerging forms of emotion display on SNS.

There are both differences and similarities between smiling flight attendants and online social networking. With the exception of a few successful bloggers, most users cannot trade an attractive profile for a wage. However, instead of profit, feelings may be managed online for identity purposes. Individualism implies a value of personal feeling and will. Given this value, Hochschild (1983/2003: 254) argues that it is worthwhile to seek out and locate one's so-called 'true feelings.' To answer the increasingly important question in late modernity of *who am I*, people turn to feelings in order to locate themselves (Hochschild 1983/2003: 22). On SNS, users then negotiate their identities and group belongings through displaying feelings. The most obvious example of this is the use of emoticons (see Dresner and Herring 2010). Punctuation and capitalization are also used online to insert feeling on SNS (Baym 2010: 103).

People have actively managed feelings in order to make their personalities fit for public contact for a long time (see Elias 1939b/1998). For example, showing how flight attendants engaged in what Hochschild (1983/2003: 36) described as 'deep acting,' their feelings were managed in such a way, that in due time, they were being experienced as true and as a part of them. Therefore in managing feelings we also contribute to their creation. This is about emotion control. My argument is that when emotions are managed and controlled in digital late modernity, the information conveyed is increasingly reflective of an ideal I and therefore about image management. I will refer to this as reflexive emotion display. Emotions online are in this way used as resources in the identity work of the user. And, as we will attend to next, instead of having a capital exchange value as for smiling flight attendants, emotion displays on SNS are rather exchanged for recognition and status.

Processes of recognition and status negotiation are made easy with digital technology and thus also clearly visible online. Messages and postings contain a form of meta-content implying that the sender is thinking about the receiver(s) (Ling and Yttri 2002: 158). Recognition is thus tied to the negotiation and maintenance of personal relationships and group belongings. And it is also tied to status. To recognize on SNS is to post on, to read, and to make links to other profiles as well as to invite to events and discussions (Hands 2011: 113). In Ling and Yttri's (2002: 159–161) study of Norwegian youth, practices of sending and receiving text messages became an objectification and quantification of popularity. In this way, to be noticed, to be recognized, was linked to status and popularity, not the least since recognition tends to be aggregative in digital late modernity. The more someone links to you, likes you, thumbs up your postings, and comments on them, etc., the higher you will be ranked and listed in the different SNS, news feeds, and tables of suggested links and readings. And since Kemper (discussed in Turner and Stets 2005: 216–220) we know that increase in status is linked to feelings of satisfaction and well-being. Indeed, positive emotions emerge when individuals are able to reaffirm their self-conceptions (Kemper, discussed in Turner and Stets 2005: 24). Collins (2004: 44) talks about 'emotional energy' in this context as a way to discuss how emotion drives behavior (or interaction rituals as he frames it), i.e., that our motivation to act is to experience positive feelings. As we shall attend to further on, such emotional energy is related to cultural capital that for a long time has had a stratifying function in societies (Bourdieu 1984/2010). Hence, emotions have a double function here, they are used to negotiate identity and status among peers online, and if successful in such negotiation, emotional energy (i.e., a feeling of confidence, elation, strength, enthusiasm, and initiative in taking action, see Collins 2004: 49) will arise which in turn can be used to reaffirm identity and status positions.

Power, Identity, and Feelings in Digital Late Modernity 23

If we conceive of identity negotiation, status, and popularity as purposes in themselves to strive for (since they give rise to positive feelings), reflexive emotion display is rational. This is a type of expressive rationality—expressing and performing identity and individuality—that is, becoming prominent in digital late modernity (see Svensson 2011). According to Elias (1969/1998: 92), such rationality was formed already in court circles of medieval Europe, evolving in conjunction with a growing sense of self-constraints. While in bourgeois (or instrumental) types of rational behavior calculation of financial gains and losses played a primary role, in court (or expressive) types of rationality, calculation of gains and losses of status are of greater importance. Hence management of feelings for status purposes has a long history. And since rationality produces forms of behavior (Elias 1969/1998: 92), it is obvious that identity negotiation through emotion display and feeling management are practices that can be traced to forms of rationalities in which prestige, status, and individuality are important.

Reflexive emotion displays online are supervised by peers in the network. Users hope that their online negotiations will be recognized by likings, linkings, postings, and thumbs up from peers. Such recognition, in turn, can be used to negotiate and reaffirm status and group belonging. Hence, we cannot reaffirm self-conceptions and negotiate emotional energy without being visible for others. At the same time, to be under the constant gaze of others has been theorized as having a disciplining effect on people (see Foucault and his treatment of the notion of the panopticon 1973/1994: 58). On SNS, users invite others to 'supervise' them in order to secure a place on the social arena, increase status and popularity. The questions then arises, when being watched by others on SNS, are users primarily being subject to the panoptic power of the gaze, or are they instead using the gaze of others to negotiate identity and status, which in turn gives rise to emotional energy? It all depends how skilfully the user navigates the SNS platforms and manages his or her databases of friends and connections on them, how tactfully the user governs his or her visibility online. To be successful, users need to master this form of sociability, through managing connections, negotiating and maintaining an attractive self in order for peers to visit, like and link their SNS profile as well as to leave comments. As we shall attend to later, some segments of the population are better equipped for doing this than others.

Elias (1969/1998: 93) claims that struggles for prestige and status can be observed in many social formations. Today SNS are examples of such social formations. Livingstone's (2008) study, for example, shows that British teenagers spend hours going from one profile to another leaving comments, something Livingstone conceives of as a necessity for these teenagers in order to reaffirm their place within the peer network. It thus seems that the increasing importance of connectivity in network societies brings with it a logic where individuality, popularity, and status, managed by reflexive emotion display and feeling management, become important ends in themselves to strive for. Thus, a network logic is a logic of reflexive emotion display, disciplining users to manage feelings (semi)publicly on SNS. This takes the form of reflexive connectivity when making links to other users public (as well as causes, organizations, brands) and hence freeloading on their supposed connotations to which users wish to tie images of themselves (Donath and boyd 2004) for the sake of enhancing their status and popularity.

Our behaviors are also influenced by norms and values in a given society (Foucault 1979/1994: 324). Such norms and values resonate in what Hochschild (1983/2003: 18) labels 'feeling rules.' Emotion management is guided by establishing a sense of entitlement and obligation (Hochschild 1983/2003: 56). When we contemplate on what we ought to feel, and when we observe how other people assess our displays of emotion, we are consciously, or unconsciously, adhering to the feeling rules in the given situation (Hochschild 1983/2003: 57, see also Elias 1939b/1998 discussion of 'shaping of feeling standards'). Here we can start to think about power in relation to emotion display and feeling management online. Disciplined practices of emotion display and (semi)public feeling management on SNS embody relations of power since such practices are closely tied to the norms and values of network society. Responsiveness, connectedness, and reflexiveness are examples of such norms and values (see Svensson 2012).

FEELING MANAGEMENT AND STRATIFICATION

In the previous section power was discussed as a kind of network logic, emerging in digital late modernity and pushing people to (semi)public reflexive emotion display among peers on SNS. This is a kind of power relation that takes form when relating to the norms and values of a given society in a given situation. In digital late modernity, it seems that reflexive identity negotiation—an urge to display a coherent and attractive self—for the sake of gaining recognition and status motivates emotion work online. Not everyone is equally skillful in attracting recognition and status. Hence, power is also at play *between* people (see also Elias 1970/1998: 115–116). In this section I will discuss which strata is more successful managing feelings and controlling displays of emotions and why this is so.

It has been argued that emotion management previously was confined to the private sphere of middle and upper classes. Referring to Kohn, Hochschild (1983/2003: 158–159) claims that middle-class kids are brought up to control and manage their feelings to a larger extent than working-class kids. The middle class is fostered in an environment in which feelings and their control are tied to power and authority (Hochschild 1983/2003). Similarly, Elias (1939a/1998: 67–69) argues for a link between feeling management and class through linking emotion control to the process of civilization. Historically the lower strata have tended to follow their emotions rather than to control them (Elias 1939a/1998). Hence, self-constraint in itself may serve as a marker of power, something that distinguishes the cool, controlled, and civilized upper classes from the affective, uncontrolled, and uncivilized lower classes (Elias 1939a/1998: 72).

In digital late modernity more segments of the population move through an increasing number of social situations and hence need to manage feelings through a wider range of different positions. Thus, the need to control emotions increases among all strata in network societies. Hence, it becomes more and more important all over the social landscape to acquire skills for managing and controlling emotions for the sake of enhancing status and popularity (see also Elias 1939a/1998: 68). What Hochschild highlights is that the ability for feeling management and emotion control differs among social classes. Therefore, future discussions on the digital divide should not only focus on the level of access to digital platforms for different groups but should also look into how successful these different groups are in achieving the goals that made them go online.

Several studies have underlined that users from lower socio-economic groups tend to be less skilled using digital platforms, hence pre-existing inequalities are both reflected and potentially increased online (DiMaggio et al. 2004; Gui and Argentin 2011; Hargittai 2008, 2010; Mossberger et al. 2003; OECD 2010). A report from the OECD (2010) concluded that the digital divide in education goes beyond access to technology, the divide is between those with the competencies and skills to benefit from computer use from those without. Such quality of use and skills is influenced by socio-economic factors (DiMaggio et al. 2004). Similarly, Hargittai and Hinnant (2008) found that among American young adults (18–26 years old), those with higher levels of education and wealthier parents used the web for more capital-enhancing activities. They incorporated the web into their everyday lives in more informed ways and for a larger number of activities (see also Hargittai 2010). Similarly, Gui and Argentin's (2011) study among high school classes in Italy shows that cultural background has a significant effect on digital skills. It thus seems that those coming from wealthier socio-economic backgrounds and with better education are more skillful managing and controlling their online activities to be used as resources.

The argument so far is that SNS are not confined to the upper and middle strata, but several studies suggest that these segments of the population are better equipped to manage them. The question that then arises is: why is this so? Here I would turn to Hsieh's (2012) treatment of the concepts of 'online social networking skills' and 'digital literacy,' i.e., how to understand and process socio-emotional meanings of various types of digital content and how to display emotions on digital platforms to reaffirm self-conceptions and negotiate status in the peer group. He discusses online social networking skills as (a) an understanding of the technological properties that enable social interactions, and (b) the knowledge of practices that increase interactivity. By this underlining of the ability to increase social interaction, the idea of online social networking skills and digital literacy can clearly be linked to practices of recognition and status in digital late modernity discussed in the previous section. Without understanding and knowing how to enable social interaction online, users will not be able to attract the necessary traffic for identity reaffirmation, status, and emotional energy.

This brings us to the question of how these differences in skills are related to educational and socio-economic backgrounds. The overarching idea here is that certain types of SNS uses can result in increased social and cultural capital (see Hargittai 2008: 940–941), which in turn can result in differentiated opportunities and life outcomes (DiMaggio et al. 2004). The highlighting of education and social origin resonates with Bourdieu's (1984/2010) treatment of cultural capital as both acquired through education and inherent in social origin (with Bourdieu emphasizing on social origin, see page 56). Such cultural capital organizes a distinction; it stratifies between groups that have it and can make use of it, and those who do not have it. And having different possible positions in social space, so-called 'habitus,' will make individuals act and relate to others and the surrounding society differently. Such habitus does not only depend on our total cultural capital and our possible positions in social space, but also in how we use this capital, since this will reveal *how* we have acquired the capital (Bourdieu 1984/2010: 58). Differences in manner and behavior indicate different modes of its acquisition (Bourdieu 1984/2010: 61). And here social origin plays a significant role. The self-confidence and ease which accompanies the certainty of possessing cultural legitimacy and a position in social space is something that is handed down in bourgeois families and cannot be taught in schools according to Bourdieu (1984/2010: 59). Hence, the natural way in which the upper and middle classes use digital platforms for enhancing status and capital can be understood by the cultural capital and habitus they have acquired and made use of.

Cultural capital and emotion have been theorized together by Collins (2004) through the concept of emotional energy, i.e., a feeling of confidence, elation, strength, enthusiasm, and initiative in taking action (see page 49). Collins (2004: 132) focuses on emotional energy instead of cultural capital, a concept he conceives as a more static and hence not as suitable for understanding stratification in micro-interactional situations. The key to understand processes of stratification in micro-interactional situations is to study inequalities in emotional energy (Collins 2004). According to Collins (2004: 133) high emotional energy gives people a kind of micro-situational legitimacy. In this way, we can connect emotional energy to status and prestige. When people feel they are in a position to understand, control, and use flows of interactions online (i.e., online social networking skills and digital literacy) they display emotional energy which tends to be returned/aggregated online in the form of status and recognition. Hence, status becomes a form of social networking capital that tends to reproduce itself. In this way flows of emotional energy among individuals are related to their status position in a network (see also Turner and Stets 2005: 215). And here we can use Bourdieu's idea of habitus to understand how the use of emotional energy is linked to different status groups. Self-perception and expectation of status and recognition in a network of peers is linked to how our positions in social space make us act and relate to each other and the surrounding society. Individuals with higher status will be given more opportunities to perform in a group (i.e., to mobilize cultural capital/ emotional energy) than individuals with lower. High status individuals will therefore receive more positive evaluations of their performances (Turner and Stets 2005: 235) and hence will be able to use this positive flow of emotional energy in their own emotion display online. In this way Bourdieu complements Collins, who treats expectation as a subconscious force anticipating emotion (see page 119) and hence does not explicitly address the link between micro-interactional situations and macro-stratification.

To sum up, in order to maintain and extend network connections and to be part of the social arena, a kind of network logic disciplines SNS users to continuous and reflexive self-revelations online. Here it seems that middle and upper strata of the population more skillfully display emotions and manage feelings for status and capital enhancing purposes. These types of online social networking skills and digital literacy can be linked to how their cultural capital and habitus provide the emotional energy in the micro-interactional situation, which in turn is related to successful reaffirmation of self-conceptions and negotiation of status positions.

EMOTION CONTROL, POWER, AND THE PROCESS OF CIVILIZATION

This chapter has put forward a critical perspective of the emerging network society through discussing power relations as constitutive of emotion display and feeling management and their connection to negotiation of personality, individuality, and status in digital late modernity. In this concluding section I will shortly discuss how Elias' theory of civilization, in which emotions and their displays are understood as socially constructed and linked to behavioral control mechanisms, can help us to think about SNS as part of the civilization process in digital late modern societies.

Increasingly rationalized displays of emotions and a growing need to manage identity online can be understood as part of a larger civilization process. According to Hochschild (1983/2003: 20), trying to manage feelings, to feel what one wants, expects, or thinks one ought to feel, is no newer than emotion itself. Hence, emotion control is an art of essential importance for civilized life (Hochschild 1983/2003: 21). Furthermore, emotion is one of culture's most powerful tools for directing action (Hochschild 1983/2003: 56). Elias is the scholar associated with feeling management as part of the civilized behavior and the broader processes of social development. Through studying how feeling standards have been shaped over generations, Elias argues that it is possible to identify a long-term civilization process. He observes how an increasing emphasis on self-control throughout history came to regulate the instinctual and affective lives of people. Volatility and mood swings that were considered relatively normal in medieval Europe were to be regarded as a form of psychological problems in later eras (Elias 1939b/1998: 55). The rise of feeling management is linked to the historical process during which the nobility was transformed from a class of knights into a class of courtiers (Elias 1939b/1998). With the stabilization of the monopoly of force, skills on the battlefield became less important compared to tactfully moderating affect and managing sentiment and conduct at the king's court (Elias 1939b/1998). According to Elias (1939b/1998: 63), this social moulding of individuals in accordance with the civilization process required a particularly intensive and stable regulation of affects.

The process of civilization led to a stricter control of affects and a higher measure of individual self-constraint (Elias 1939a/1998: 71). Elias (1939b/1998: 58) understands constant self-control as the management of feeling and moderation of affective charges in our self-expressions. Such processes are linked to identity and personality since practices of self-control form a part of the personality structure of the individual (Elias 1939b/1998: 58). Hence, to reflect upon one's personality and how and what emotion should be displayed, is to control oneself (see Elias 1987/1998: 288). And such control of behavior has been used to access power in the history of the civilization process (Elias 1969/1998: 92). Hence, with Elias' theorizing of the civilization process, practices of feeling management and the importance of identity negotiation are also tied to historical and societal developments of increasing social interdependence.

What happened during the course of civilization was that emotional expression was increasingly controlled and emotions thus became more reliable and calculable (Elias 1939b/1998: 49). People's social functions were more and more differentiated, hence people became increasingly dependent on each other (Elias 1939b/1998: 51–52). Therefore a need to attune conducts and emotion displays more strictly and accurately arose in order for people to fulfill their social functions (Elias 1939b/1998). Thus, the interdependence of people corresponds with the increasing importance of feeling management during the course of civilization. If we apply Elias' argument in our digital and late modern age, social functions continue to be differentiated. People are even more interdependent in connected network societies. Hence, there will be a continuous need to control visibility and emotion display in order to manage identity, personality, and sociability. Social networking online could be understood in light of this need for a personal control system. In other words, the process of civilization continues online.

Power, Identity, and Feelings in Digital Late Modernity 29

The importance of human interdependence together with rationalized practices of affect moderation and management of feelings and conduct highlight how power is linked to reflexivity. Elias does not discuss reflexivity explicitly, but he uses related concepts such as the "interweaving of individuals when negotiating personality" (see 1987/1998: 274). Here Elias (1987/1998: 275) connects individualization with the increasingly dominant perception of humans as closed systems in modernity (so-called 'homo clausus'). The idea of the individual as a closed system brought with it a possibility to conceive of yourself from a distance. In other words, to be reflexively self-conscious is to be individualized (Elias 1987/1998: 276). By discussing this as part of the civilization process, reflexivity is tied to power since such reflexivity brings with it practices of self-restraint and control, a greater need to observe and think before acting (Elias 1987/1998: 280). Here the late modern concept of reflexivity, usually understood as liberating the individual from the institutions of modernity (see Beck 1995; Giddens 1991), becomes part of a personal control system online. According to Elias (1987/1998: 279), this brings with it a double role for people as both observers and observed. This double role is clearly accentuated on SNS. In this way Elias' theory of the civilization process is still valuable for understanding power in contemporary connected and late modern societies.

In conclusion, what is the rationality of emotion display online? By understanding SNS as personal control systems for monitoring and controlling individuality, managing feeling, and displaying emotion in front of selected others, late modern reflexive individualization is connected to both micro-relations of power of status negotiation through emotional energy, and to macro-relations of power within processes of stratification and civilization. To reaffirm self-conceptions by displaying emotions online and monitor how they are received is to attune conduct to a network society of increasing interdependence and connectedness. The rise of (semi)public management of feelings can thus be understood in light of an increasing need to account for others and co-exist in a society where connections between people multiply and connections in themselves become more and more important. Hence the purpose of feeling management on SNS is social as well as individual and also part of the process of civilization. By understanding emotion displays online in this way, the values and the norms of connectedness, responsiveness, and reflexive individualization are brought to the fore in network societies. Relations of power between people in network societies will increasingly revolve around whose displays of emotion will be successful in enhancing cultural capital and emotional energy and making use of it. These conclusions will benefit from future empirical studies on how the values and norms of connectedness, responsiveness, and reflexive individualization push/discipline people to reflexive emotion displays online, and exactly how these displays of emotional energy are used to negotiate status and positions in the peer network.

ACKNOWLEDGMENTS

This article has been made possible with support from the Wahlgrenska Foundation, HumanIT, and Karlstad University.

REFERENCES

Bakir, V. (2010) *Sousveillance, Media and Strategic Political Communication. Iraq, USA, UK*. New York: Continuum.
Baym, N. (2010) *Personal Connections in the Digital Age*. Cambridge: Polity Press.
Beck, U. (1995) *Att uppfinna det politiska; Bidrag till en teori om reflexiv modernisering* [To Invent Politics. Contribution to a Theory of Reflexive Modernization]. Göteborg: Daidalos.
Benkler, Y. (2006) *The Wealth of Networks: How Social Production Transforms Markets and Freedom*. New Haven, CT: Yale University Press.
Bourdieu, P. (1984/2010) *Distinction*. London: Routledge.
Bruns, A. (2008) *Blogs, Wikipedia, Second Life, and Beyond. From Production to Produsage*. New York: Peter Lang.
Castells, M. (2000) *Nätverkssamhällets framväxt. Informationsåldern. Ekonomi, samhälle och kultur. Band 1* [The Rise of the Network Society. The Information Age: Economy, Society and Culture, Vol. I] (2nd ed.). Göteborg: Daidalos.
Castells, M. (2001) *The Internet Galaxy: Reflections on the Internet, Business and Society*. Oxford: Oxford University Press.
Collins, R. (2004) *Interaction Ritual Chains*. Princeton, NJ: Princeton University Press.
Dahlgren, P. (2009) *Media and Political Engagement. Citizens, Communication, and Democracy*. New York: Cambridge University Press.
DiMaggio, P., Hargittai, E., Celeste, C. and Shafer, S. (2004) Digital Inequality: From Unequal Access to Differentiated Use. In. K.M. Neckerman (Ed.), *Social Inequality*. New York: Russell Sage Foundation, pp 355–400.
Donath, J. and boyd, d. (2004) Public Displays of Connection. *BT Technology Journal*, 22(4), 71–82. Retrieved from http://www.danah.org/papers/PublicDisplays.pdf, March 14, 2009.
Dresner, E. and Herring, S. (2010) Functions of the Nonverbal in CMC: Emoticons and Illocutionary Force. *Communication Theory*, 20(3), 249–268.
Elias, N. (1939a/1998) Diminishing Contrast, Increasing Varieties. In S. Mennell and J. Goudsblom (Eds.), *Norbert Elias. On Civilization, Power and Knowledge*. Chicago: The University of Chicago Press, pp 67–74.
Elias, N. (1939b/1998) The Social Constraint Towards Self-Constraint. In S. Mennell and J. Goudsblom (Eds.), *Norbert Elias. On Civilization, Power and Knowledge*. Chicago: The University of Chicago Press, pp 49–66.
Elias, N. (1969/1998) The Changing Functions of Etiquette. In S. Mennell and J. Goudsblom (Eds.), *Norbert Elias. On Civilization, Power and Knowledge*. Chicago: The University of Chicago Press, pp 83–94.
Elias, N. (1970/1998) Game Models. In S. Mennell and J. Goudsblom (Eds.), *Norbert Elias. On Civilization, Power and Knowledge*. Chicago: The University of Chicago Press, pp 113–138.
Elias, N. (1987/1998) Homo Clausus: The Thinking Statues. In S. Mennell and J. Goudsblom (Eds.), *Norbert Elias. On Civilization, Power and Knowledge*. Chicago: The University of Chicago Press, pp 269–290.

Ellison, N. and boyd, d. (2007) Social Network Sites: Definition, History and Scholarship. *Computer-Mediated Communication*, 13(1), Article 11. Retrieved from http://jcmc.indiana.edu/vol13/issue1/boyd.ellison.html, February 5, 2009.
Foucault, M. (1973/1994) Truth and Juridical Forms. In J.D. Faubion (Ed.), *Power—Essential Works of Foucault 1954–1984, Volume 3*. London: Penguin Books, pp 1–89.
Foucault, M. (1979/1994) Omnes et Singulatim. In J.D. Faubion (Ed.), *Power—Essential Works of Foucault 1954–1984, Volume 3*. London: Penguin Books, pp 298–325.
Giddens, A. (1991) *Modernity and Self-Identity: Self and Society in the Late Modern Age*. Cambridge: Polity Press.
Gui, M. and Argentin, G. (2011) Digital Skills of Internet Natives: Different Forms of Digital Literacy in a Random Sample of Northern Italian High School Students. *New Media and Society*, 13(6), 963–980.
Hands, J. (2011) *@ is for Activism. Dissent, Resistance and Rebellion in a Digital Culture*. New York: PlutoPress.
Hargittai, E. (2008) The Digital Reproduction of Inequality. In. D.B. Grusky in collaboration with M.C. Ku and S. Szelényi (Eds.), *Social Stratification: Class, Race, and Gender in Sociological Perspective*. Boulder: Westview Press, pp 936–944.
Hargittai, E. (2010) Digital Na(t)ives? Variation in Internet Skills and Uses Among Members of the "Net Generation." *Sociological Inquiry*, 80(1), 92–113.
Hargittai, E. and Hinnant, A. (2008) Digital Inequality. Differences in Young Adults Uses of the Internet. *Communication Research*, 35(5), 602–621.
Hochschild, A.R. (1983/2003) *The Managed Heart. Commercialization of Human Feeling*. Berkeley, CA: University of California Press.
Hsieh, Y.P. (2012) Online Social Networking Skills: The Social Affordances Approach to Digital Inequality. *First Monday*, 17(4). Retrieved from http://firstmonday.org/htbin/cgiwrap/bin/ojs/index.php/fm/article/viewArticle/3893, June 10, 2012.
Katz, J.E. and Aakhus, M. (2002) *Perpetual Contact. Mobile Communication, Private Talk, Public Performance*. Cambridge: Cambridge University Press.
Leaning, M. (2009) *The Internet, Power and Society: Rethinking the Power of the Internet to Change Lives*. Oxford: Chandos.
Ling, R. and Yttri, B. (2002) Hype-Coordination Via Mobile Phones in Norway. In J.E. Katz and M. Aakhus (Eds.), *Perpetual Contact. Mobile Communication, Private Talk, Public Performance*. Cambridge: Cambridge University Press, pp 139–169.
Livingstone, S. (2008) Taking Risky Opportunities in Youthful Content Creation: Teenagers' Use of Social Network Sites for Intimacy, Privacy and Self-Expression. *New Media and Society*, 10(3), 393–411.
Medierådet. (2010) *Ungar and Medier 2010. Fakta om barns och ungas användning och upplevelser av medier* [Youth and Media 2010. Facts About Children's and Youths' Use and Perceptions of Media]. Stockholm: Kulturdepartementet. Retrieved from http://www.medieradet.se/upload/Rapporter_pdf/Ungar%20och%20medier%202010.pdf, November 2, 2010.
Miller, V. (2008) New Media, Networking and Phatic Culture. *Convergence The International Journal of Research into New Media Technologies*, 14(4), 387–400.
Mossberger, K., Tolbert, C.J. and Stansbury, M. (2003) *Virtual Inequality: Beyond the Digital Divide*. Washington, DC: Georgetown University Press.
OECD. (2010, March) *Are New Millennium Learners Making the Grade? Technology Use and Educational Performance in PISA 2006*. Retrieved from http://www.oecd-ilibrary.org/education/are-the-new-millennium-learners-making-the-grade_9789264076044-en, May 16, 2012.

O'Reilly, T. (2005) *What is Web 2.0. Design Patterns and Business Models for the Next Generation of Software*. Retrieved from oreilly.com/web2/archive/what-is-web-20.html, February 5, 2009.

Pröitz, L. (2007) Mobile Media and Genres of the Self. In T. Storsul and D. Stuedahl (Eds.), *Ambivalence Towards Convergence: Digitalization and Media Change*. Göteborg: Nordicom, pp 199–216.

Rheingold, H. (2002) *Smart Mobs. The Next Social Revolution. Transforming Cultures and Communities in the Age of Instant Access*. Cambridge: Basic Books.

Rice, R.E. and Love, G. (1987) Electronic Emotion: Socioemotional Content in a Computer-Mediated Communication Network. *Communication Research*, 14(1), 85–108.

Shirky, C. (2009) *Here Comes Everybody. How Change Happen When People Come Together*. London: Penguin Books Ltd.

Svensson, J. (2011) The Expressive Turn of Citizenship Digital Late Modernity. *JeDem—eJournal of eDemocracy*, 3(1), 42–56.

Svensson, J. (2012) Social Media and the Disciplining of Visibility. Activist Participation and Relations of Power in Network Societies. *European Journal of E-Practice*, 16, 16–28.

Turner J.H. and Stets, J.E. (2005) *The Sociology of Emotions*. Cambridge: Cambridge University Press.

van Dijk, J. (2006) *The Network Society* (2nd ed.). London: Sage.

Walther, J.B, van der Heide, B., Kim, S., Westerman, D. and Tong, S.T. (2008) The Role of Friends' Appearance and Behavior on Evaluations of Individuals on Facebook: Are We Known by the Company We Keep? *Human Condition Research*, 34. Retrieved from http://onlinelibrary.wiley.com/doi/10.1111/j.1468-2958-.2007.00312.x/full, February 5, 2009.

2 Feeling Through Presence
Toward a Theory of Interaction Rituals and Parasociality in Online Social Worlds

David Boyns and Daniele Loprieno

INTRODUCTION

The study of parasocial relationships (Horton and Wohl 1956) has been a prominent concern for investigations into mediated interaction for several decades. Of notable interest are the emotional qualities that parasocial relationships engender, and the importance of emotions in generating feelings of social connectedness in virtual environments. This chapter is a theoretical examination of the Internet as a parasocial environment in which emotional interactions are prominent in social interaction. The discussion begins with a review of Interaction Ritual Theory (IRT) developed by Randall Collins (1981, 2004), and its relevance for the study of online interaction. It then moves to an analysis of the literature on parasociality (Horton and Wohl 1956) and its relationship to the study of virtual social worlds. We argue that research into parasocial interaction provides an excellent foundation for a re-examination of IRT. Although IRT limits itself by emphasizing face-to-face (F2F) interaction, our argument expands Collins' IRT by examining the factor of "co-presence" within Internet-based, parasocial relationships. We review the growing literature on "presence" (Lombard and Ditton 1997) that has investigated the ability of virtual environments to simulate co-presence and create parasocially shared emotional experiences. We also examine the role of sociological context in the formation of parasocial co-presence and the means through which online emotional experiences can be translated into real–space relationships. To support the theoretical ideas developed within the chapter, research from online communities, chat-rooms, and virtual worlds are examined. Finally, the chapter discusses the theoretical implications of examining IRT in the context of the Internet, and proposes that interaction rituals (IRs) in both real and virtual spaces intersect to produce an unprecedented, emotional experience for individuals.

THE INTERNET AND INTERACTION RITUAL THEORY

Emotions have become a central concern of research into online environments. Recent investigations have suggested that emotional experiences are a prominent, if not crucial, aspect of the engagement of individuals in online worlds (Ben-Ze'ev 2004; Boyns 2010; Derks et al., 2008; Turkle 2011). The growth of interest in the study of emotions within virtual worlds has paralleled the increasing sophistication and prevalence of online technologies, and has also run tandem to the development of a coherent interest in the sociology of emotions. The conjunction of these two areas of study provides fertile ground for the study of emotions in online environments, and forms the basis for our argument.

Interaction Ritual Theory

It is only recently that sociology has begun to systematically incorporate emotions into the analysis of human social life (Turner and Stets 2005). Until the 1970s, emotions were marginalized in sociological analysis, but since, have become a substantively rich dimension of the study of social interaction. The tradition of theory that has most directly incorporated emotions as a primary component of social interaction has been IRT developed by Randall Collins (1981, 2004). While Collins' IRT is well-known it has not been thoroughly examined in the context of technologically-mediated interaction (TMI) (see Ling 2008 for an example). Collins (2011) does note that one of the most prominent questions posed about IRT is the role of electronic media in creating emotional energy; however, Collins is dubious about the efficacy of TMI in producing successful IRs.

Collins' theory extends Durkheim's (1912/1995) analysis of ritual and emotion in social experience, and, following Goffman (1967), situates it into the micro-dimensions of everyday IRs. Collins argues that the core of social life is the transmission of "emotional energy" through interaction, facilitating the integration of members of a society, and creating the common denominator for social action (Collins 1993). He suggests that chains of IRs provide the microfoundation of social life and are the primary means by which the social world becomes "charged" with emotions. For Collins, IRs have four primary ingredients: physical co-presence, common focus of attention, shared emotional mood, and a boundary to outsiders. When effectively combined, these ingredients result in the production of an "emotional energy" that provides members of the interaction with a sense of efficacy, group membership, and the emotional bonds of solidarity. Participants in the IR seek to memorialize this emotionally energized feeling by representing it with symbols (e.g., group icons, family photos, marks of status, etc.) that can be utilized in recalling the positive feelings of the collective experience. When invoked, these symbols can reanimate emotional energy, revive interpersonal bonds, and provide a foundation for self-efficacy and creativity (Collins 2004). The invocation of symbols is important in IRT because emotional energy is an interpersonally produced resource that decays through time and must be recharged through social interaction (Collins 2004; Hanneman and Collins 1998). Collins (1981, 2004) argues that the emotional energy attached to symbols is fundamental for social life, the primary source of social solidarity, the driving motivation toward interaction, and, ultimately, the glue that binds together social networks. Collins' IRT is illustrated in Figure 2.1.

Because of its emphasis on emotions in social interaction, Collins' IRT provides an important foundation for an analysis of TMI. However, Collins is reluctant to extend the IR process into non-F2F interaction. He argues that the more technologically mediated our forms of interaction become, the less likely they are to generate emotional energy and produce feelings of solidarity. In reflecting on the Internet's ability to produce effective IRs, Collins' (2004: 64) theoretical position is clear: "The more that human social activities are carried out by distance media, at low levels of IR intensity, the less solidarity people will feel; the less respect they will have for shared symbolic objects; and the less enthusiastic personal motivations they will have in the form of EE [emotional energy]." For Collins, because bodily presence is central in shaping the focus of attention and mutual engrossment required for a successful IR, he contends that it is unlikely that TMI can create potent, emotional content. He summarizes the possibilities of online IRs by suggesting that "the hypothesis of IR theory is that face-to-face communication will not disappear in the future, nor will people have any great desire to substitute electronic communications for bodily presence" (Collins 2004: 63).

For Collins, it is the primacy of "bodily presence" that anchors IRs in the social experience. Without physical co-presence, IRs can only produce attenuated forms of emotional energy and concomitantly low levels of social connection. This emphasis on co-presence is perhaps an important limitation of Collins' IRT, especially considering the prevalence of electronic communication in conducting daily interaction (Ling 2008). Despite this limitation (and perhaps because of it), Collins' IRT lends itself nicely to the study of online interaction. While participants in TMIs commonly encounter many of the same ingredients found in IRs (i.e., they often focus on common topics, generate a common mood, and create a psycho-social

Interaction Ritual Ingredients
- Boundary to Outsiders
- Common Focus of Attention
- Shared Mood
- Group Assembly (Bodily Co-Presence)

→

Interaction Ritual Outcomes
- Emotional Energy
- Group Solidarity
- Sacred Symbols
- Group Identity

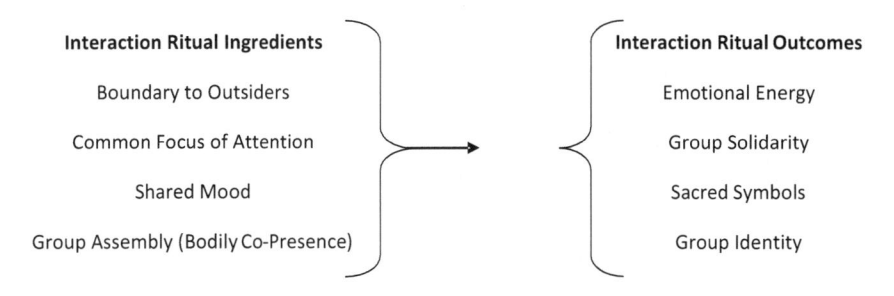

Figure 2.1 Dynamics of interaction rituals.
(Adapted from Collins, 1981, 2004a)

boundary separating themselves from external social conditions) the one ingredient that is not shared is physical co-presence. Following the model of IRT, the central concern for the study of online communication becomes the degree to which forms of TMI can transcend F2F bodily presence and create simulations of co-presence. It is this concern that frames the analysis of this chapter.

EMOTIONS, PARASOCIAL INTERACTION, AND CONVERGENCE CULTURE

The study of the sociological dimensions of TMI is not altogether new. In fact, research in this area has evolved over several decades and in a broad range of subject areas (e.g., online role-playing games, social relations with media figures, fan cultures, and online communities). An important subset of this literature has developed around the seminal concept of "parasocial interaction" (Horton and Wohl 1956), a form of social interaction that is accomplished at a distance, primarily through mass media, and is characterized by the creation of an "illusion of intimacy" with mediated personas. In parasocial interaction, communication unfolds without F2F presence, but it emerges as a "simulacrum of conversational give and take" (Horton and Wohl 1956: 216) between users and their technologically-mediated others. Sometimes these mediated others are imaginary characters from the fictional realms of cinematic, televisual, and literary media (e.g., characters in soap operas, comic books, films); at other times they are persons of real-world existence whose presence is simulated through mediated technology (e.g., pop stars, sports heroes, celebrities). Horton and Wohl (1956) argue that continued exposure to media figures allows viewers to "know" mediated characters much in the same way that they might "know" their friends. This feeling of "knowing" a mediated figure allows the spectator to treat the fictional or distant as real and intimate. As this connection is sustained over time, it can take on the qualities of sociability much like that experienced in intimate, personal relationships (Alperstein 1991; Auter 1992; Ballantine and Martin 2005; Canary and Spitzberg 1993; Caughey 1984; Cohen 1997, 2004; Eyal and Cohen 2006; Leimeister et al. 2008; Rubin and Perse 1987).

While researchers have traditionally explored parasociality in the context of broadcast media (e.g., television and radio), research over the past several decades has extended the concept of parasocial interaction into the study of online worlds (Ballantine and Martin 2005). Much of the interest in the parasocial nature of online worlds has been derived from two sources: the study of computers, themselves, as social others (Lee 2004; Lee and Nass 2005; Nass and Moon 2000; Reeves and Nass 1996); and investigations of the communities that form in online environments (for early examples see Dibbell 1998; Jones 1995; Rheingold 1993; Turkle 1995).

When compared with broadcast media, there are unique parasocial aspects to the human–technology interface that emerge among the participants in virtual communities. The most notable distinction is that computer interaction as a communications medium is more parasocially interactive, resulting in the evolution of normatively organized, "participatory cultures" created by users (Hardey 2008; Nardi 2010; Nieckarz 2005; Raessens 2005). Accounts of life in virtual worlds reveal the power of online technologies to establish romantic relationships (Attwood 2009; Ben-Ze'ev 2004; Hardey 2002), create meaningful "virtual identities" (Dibbell 1998; Turkle 1995; Kendall 1998; Nieckarz 2005), and develop friendships (Rheingold 1993; Henderson and Gilding 2004).

Though online environments increasingly exist as sociologically rich social spaces, it is important to note that they emerge in a cultural context where they are increasingly interrelated with experiences off the screen (Castronova 2005, 2007; Chayko 2002; Coleman 2010; Hardey 2008; Reeves and Nass 1996). As Jenkins (2008) has argued, we are witnessing a "convergence culture" where online media have created opportunities for a variety of media platforms to intersect, creating new and complex contexts for social interaction, and providing opportunities for the emergence of rich participatory experiences. In the contemporary, hyper-mediated environment, parasocial encounters develop in a complex interface of technologically mediated and real-life interactions. In the context of such developments, it is clear that Internet-based parasocial relations are increasingly playing a more prominent role in interpersonal communication, and they surely have consequences for the dynamics of IRs. One might argue that mediated and real-life interaction have become so thoroughly conjoined that it is no longer practical to make clear distinctions between them.

RETHINKING INTERACTION RITUALS IN THE CONTEXT OF PARASOCIAL INTERACTION

Using the framework of parasocial interaction, we examine a specific aspect of IRT: the emphasis on co-presence. While Collins (2004) argues that the efficacy of IRs is attenuated in communications media, a growing literature suggests that individuals can develop highly salient socio-emotional relationships through online interaction (Banos et al. 2004; Bente et al. 2008; Hai-Jew 2009; Riva et al. 2007; Tan 2008). We examine the degree to which forms of parasocial interaction can create simulations of co-presence that transcend the IRT limitations of bodily presence. The research into what has become known as "presence" and "parasocial presence" has taken the forefront of this discussion and has illuminated the means through which co-presence can be technologically simulated.

We summarize our argument in Figure 2.2. Here, we suggest that the experience of "parasocial presence" is derived from what we describe as

Interaction Ritual Ingredients

Boundary to Outsiders

Common Focus of Attention

Shared Mood

Group Assembly (Bodily Co-Presence)

Parasocial Simulation of Co-Presence

Emotional Immersion

Parasocial Presence

"Illusion of Intimacy"
(Parasocial interaction)

"Illusion of Non-Mediation"
(Social Presence)

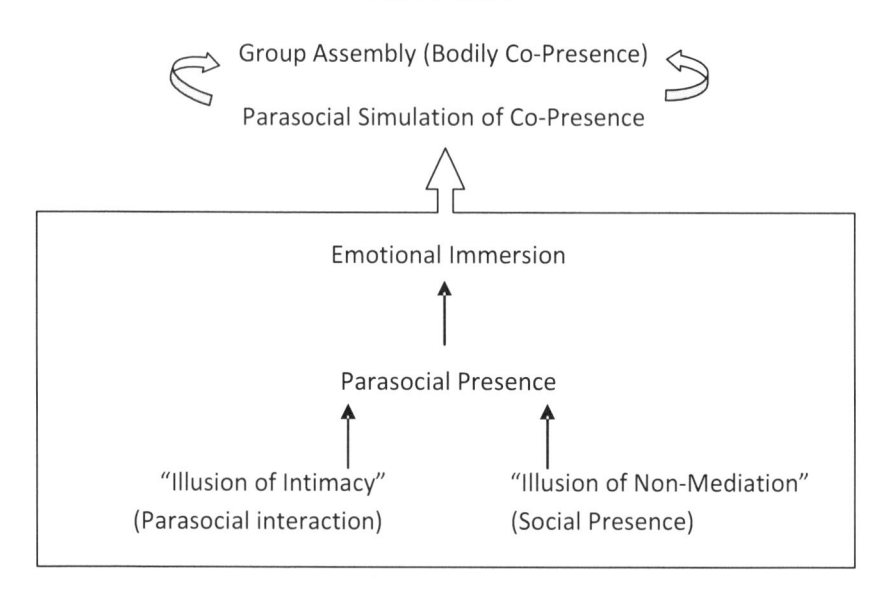

Figure 2.2 Dynamics of parasocial interaction rituals.

two "illusions" produced by TMI: the "illusion of intimacy" described in the literature on parasocial interaction, and the "illusion of non-mediation" outlined in the literature on social presence. Through experiences of "parasocial presence," we suggest that individuals become emotionally immersed in TMI, and their interactions develop qualities that simulate the experience of co-presence and take on the more general characteristics of IRs. We spell out the details of this argument in the following sections.

Interaction Rituals and the "Presence" of Co-Presence

The theory of "presence" has developed largely as a result of research into the interface between human physiology and communications technology, and has been motivated by the growing interest in virtual reality. Research into presence has directly explored the technological simulation of co-presence, and provides important theoretical insight into the factors that are conducive to producing the parasocial dynamics of IRs. Presence has been defined as "a mediated experience that seems very much like it is not mediated ... [it] creates for the user a strong sense of presence"; or more simply, the hallmark of presence is "the perceptual illusion of nonmediation" (Lombard and Ditton 1997). While studies of parasociality have investigated the means through which TMI can create an "illusion of intimacy," the presence literature has pushed this discussion further in examining how these technologies can facilitate the "illusion of nonmediation." It is our contention that a synthesis of these approaches can help to understand the ways in which TMIs can produce parasocial IRs that are experienced as if they were both "intimate" and "nonmediated."

Lombard and Ditton (1997) review a wide range of research efforts into presence and directly explore the degree to which media technologies are able to produce a simulated sense of F2F presence. The results of their review suggest that traditional technologies like print, telephone, radio, and television have typically provided a limited sense of presence because they utilize a narrow number of sensory modalities. However, the technological evolution of communications media has invigorated the possibilities of presence—e.g., computer-generated animation, 3D technology, virtual reality gear, smart phones, multi-modal Internet communication, real-time video interaction, etc. The sum of these technologies has served to provide a basis whereby TMI can more effectively replicate a felt sense of authentic, F2F interaction.

Lombard and Ditton argue that there are three important sets of variables in creating conditions of presence: formal variables, content features, and the characteristics of media users. Conspicuously missing from Lombard and Ditton's theory are sociological factors that facilitate presence. Research into what is known as "*social* presence" (Bente et al. 2008; Lee and Nass 2005; Nowak and Biocca 2003) has begun to address this gap. Here, social presence is understood as the quality of interaction that occurs when TMI creates the impression that other individuals are socially present (Lee 2004; Lee and Nass 2005). Ostensibly, these individuals can have a real-world existence outside of the media (e.g., as in phone conversations, or personal emails); they can be present only parasocially (e.g., as in spam marketing messages, or television soap operas); or they can exist as pure technological simulations of social others (e.g., as in chat-room bots, or even as computers themselves). While these technologically-mediated others can be real or imaginary, it is argued that the key concern in creating a felt sense of presence is the nature and extent of "people's *social* responses to media" (Lee and Nass 2005: 33, emphasis added). The general conclusion reached by this perspective is that if technologically-mediated environments appear to be inhabited by entities with human and social characteristics these environments can create social responses to media that reflect a felt sense of presence.

The presence and social presence research suggest that through TMIs individuals can experience a *nonmediated* presence of others in the absence of bodily presence. However, this conclusion does not also suggest that these individuals develop feelings of *intimacy* that are examined within the literature on parasocial interaction. Recent efforts to synthesize these two literatures have emerged in examinations of online environments and their ability to produce "parasocial presence" (Kumar and Benbasat 2002: 5), or technologically-mediated experiences that facilitate "a sense of understanding, connection, involvement and interaction." The concept of parasocial presence explores those TMIs that create both a sense of intimacy (i.e., "understanding, connection and involvement") and nonmediation ("interaction"), and it does so by developing a complementary synthesis of the research on parasocial interaction and social presence. Thus, parasocial presence allows for a multiplicity of interactions to be characterized as having a parasocial quality, whether through synchronous interactions like video conferencing, instant messaging, and avatar interaction, or through asynchronous modalities like text messaging, email, forum posting, eBay auctions, and website browsing.

While little research has been done to specifically examine the concept of parasocial presence, investigations in the area of e-commerce have suggested that websites that provide parasocial presence do create stronger feelings of trust and loyalty among users (Teoh and Cyril 2008). This literature joins the growing body of presence literature in revealing that parasocial interaction can facilitate a felt sense of presence particularly when individuals have accompanying experiences of emotional connection (Banos et al. 2004; Bente et al. 2008; Riva et al. 2007; Tan 2008). Do these parasocial experiences of emotional connection rival those produced by F2F interaction? On this point, the general results of these studies are not entirely conclusive. However, these studies do provide evidence that TMIs can create feelings of emotional connection matched with a felt sense of presence (Banos et al. 2004; Hai-Jew 2009).

Emotions, Immersion and Co-Presence in Online Worlds

While the body of research into presence may not provide conclusive evidence for the efficacy of IRs in virtual social environments, studies of online worlds are beginning to provide compelling accounts of the power of the Internet to create emotionally-engrossing and immersive experiences. Research into online worlds (e.g., Second Life, World of Warcraft, EverQuest, and other Massively Multiplayer Online Role Playing Games [MMOs]) has begun to document individuals' rich immersion in technologically-mediated environments. Because online worlds, like MMOs, are able to create rich immersive experiences, they seem to be able to translate a parasocial experience into a felt sense of intimate co-presence.

The idea of immersion in online worlds is central to the game design literature (Bartle 2004) and describes the experiences individuals have of feeling that they are genuinely "inside" a virtual environment. Immersion is a sought-after experience by online game designers, and when gaming environments are successful users frequently describe the immersive quality of the game as being a primary motivating factor for their game-play (Yee 2006). Recent research into online worlds has applied Huizinga's (1950) idea of the "magic circle" to the experiential dimensions of virtual games (Bainbridge 2010; Castronova 2005; Juul 2005; Nardi 2010; Salen and Zimmerman 2003), and joins a growing literature that suggests that inhabiting the "magic circle" of a virtual space can develop into a truly engrossing and emotionally rich experience.

Huizinga's notion of the "magic circle" can help us to compare the dynamics of online interaction with those of F2F IRs. The idea of a "magic circle" is a metaphor for the unique social boundaries that emerge around individuals who share a common activity, a common set of rules, and the understanding that they inhabit a collective social space. But, it is not only games that are enclosed by "magic circles"; online interactions, email exchanges, individuals posting on forums, or even viewers of online videos and webpages can also be surrounded by their own circles. In fact, all interactions can be said to be similarly enclosed, even F2F interactions, and this is the central premise of Collins' IRT: successful IRs encircle, mutually attune, and separate individuals from other social spaces, even if only temporarily; the result is an emotionally energized and unifying experience among the participants. While TMIs do produce limitations on co-presence, the communities that emerge within these worlds can have emotionally energized interactions, such that their participants can be made to feel co-present, even if only temporarily. Much like Collins' analysis of IRs, we would expect that as online IRs iterate through time, they work to generate solidarity among their participants, and will be more likely to produce and sustain emotional energy.

Research that investigates the iterative nature of online interaction has also demonstrated that not only do online interactions create emotional bonds but they also create richer emotional experiences. For example, studies of fan communities have demonstrated that interaction in online forums can create parasocial relationships with celebrities (Kassing and Sanderson 2009). Fans may draw inspiration from these relationships and come to believe that they interact with their idols in intimate ways (e.g., celebrating victories, offering sympathy, collectively offering advice, etc.). Such research is indicative of the parasocial relationships the Internet can create between fans and celebrities, and among fans themselves. Because many of these online forums develop through instantaneous updates, fans often feel as if they are present with the celebrities, sharing their moment-by-moment experiences, and cultivating emotional reactions to victories, humiliations, and struggles (Kassing and Sanderson 2009).

The iterative nature of many online interactions has also yielded important consequences for creating a sense of co-presence among participants. Studies of online relationships illustrate that many individuals experience online environments as "real" social spaces with "real" inhabitants. The emphasis on the perceived "reality" of these parasocial places is essential to how individuals experience others online and the type of relationships that are formed. For example, studies of eDating (Hardey 2008) reveal that users develop immersive, parasocial relationships when they make meaningful and trusting connections with others, and identify online environments as "real places." Similar studies of online sex chat-rooms also illustrate that those participants who perceive interactions with other users as "real" also define their sexual interactions as "real sex" and have richer emotional experiences (Attwood 2009). The results of these studies suggest that there does appear to be a connection between immersion in online interaction, where distinctions between the "real" and the "virtual" are blurred, and individuals' tendencies to experience a sense of co-presence and emotional connection in their relationships.

Intersections of F2F and Parasocial Interaction

Concerns over the perceived "reality" of online interaction draws on another issue related to the online dynamics of IRs. The parasocial dynamics of online interaction are not simply a function of the interface between individuals and the technologies with which they interact. The human engagement of technologically-mediated worlds is a sociological experience that occurs within these worlds, and also outside of them. A sociological examination of virtual worlds suggests that the reality that individuals experience and construct is a product of multiple levels of engagement (Boyns et al. 2009), and their immersion in these worlds is a function of interactions that occur on both sides of the technologically-mediated veil (Schiano et al. 2011).

While TMIs take place parasocially, with varying degrees of "realism," they also exist in a context of real-world interactions and other parasocial interactions. Individuals may be immersed in the "magic circle" of an online community, but they also talk about, and even reinforce, this group outside of the circle. In some cases, they simply shift the "circle" of the community by moving their interactions into a different medium, like when the players of an MMO discuss their game-play on an online forum, or meet in real-life to outline strategy (see Boyns et al. 2009; Nardi 2010, Taylor 2006). In other cases, individuals talk about their online interactions with entirely different groups of individuals, as many people do when they are emotionally energized. What happens in one community, interaction, or medium does not exist in isolation from other communities, interactions, and media; and both parasocial and F2F worlds frequently intersect (for examples see Boyns et al. 2009; Brown 1994; Harrington and Bielby 1995; Taylor 2006). Following Collins, we suggest that the movement of IRs betwixt and between the veils of real and parasocial worlds create a complex social arena through which IRs can flow.

Studies into online relationships have begun to examine the intersection of online and offline interaction, and reveal that F2F interaction has a significant impact on online relationships, and vice versa. There are many examples of the interpenetration of offline and online worlds in the study of MMOs (Boyns et al. 2009; Nardi 2010; Taylor 2006). Perhaps the most poignant, interpersonal examples for the study of IRT come from studies of the rich emotional relationships that emerge in online dating. Such research suggests that online chat is an essential component of offline IRs, where individuals utilize successful and repeated online communication to evaluate whether or not potential mates are worth pursing offline (Hardey 2008). The ability of potential dating partners to communicate emotionally online can be essential for conveying needs and insecurities prior to an offline meeting (Hardey 2008). Similarly, for men in sex chat-rooms, those who engage in offline meetings are more likely to describe the online interactions as real, to indicate an emotional connection with partners from the site, and to describe the chat-room as a type of community (Attwood 2009).

Similar results have been identified in research on online communities that meet offline during community conventions (Coleman 2010; Taylor 2006). For example, Coleman's (2010) account of a hackers' conference suggests that individuals meeting for the first time draw on their repertoire of online interactions to create offline relationships, and that these offline connections exist primarily to reaffirm their online affiliations. Utilizing Collins' IRT to frame the experience of individuals attending these conferences, Coleman (2010) explains that offline encounters between hackers are the culmination of sustained parasocial interactions between the members of the group. While the conference may embody the energized and emotional dynamics of IRs, the central purpose of the conference for many individuals is to strengthen their online interpersonal bonds. Coleman (2010: 68) closes her observations noting that, "As one sits at her computer, coding feverishly for a project, thousands of miles away from some of her closest friends . . . one has to wonder, 'does this matter to others in the same way as it does to me?' . . . the hacker conference answers with lucidity and clarity."

CONCLUSION

This chapter has examined Collins' IRT in the context of online environments. While Collins' argument describes the limitations of Internet interaction on IRs, our analysis contends that IRT can be extended into online environments. Individuals do seem to become emotionally engrossed and immersed in online worlds. In fact, our analysis suggests that IRT stands to benefit significantly by the growing research into parasociality and parasocial presence. If online IRs can be seen to reflect similar outcomes to those F2F, namely the generation of emotional energy, then perhaps there is room for an expansion of not only the empirical applicability but, also, the theoretical scope of IRT.

In a contemporary context of increased moment-to-moment use of electronic media, IRs across different media are likely to be chained together. The continuity of conversations, relationships, and communities are likely to be extended through both F2F interactions, and parasocial interactions. Such dynamics also raise the possibility of IRs that are purely parasocial in nature. If individuals can extend, and perhaps develop, emotional energy with others through parasocial means, perhaps these others need not have real-world existence, and can be simply the product of simulations of co-presence. Such possibilities point to important directions for theory and research in extending IRT into the study of online environments.

The combination of F2F and online IRs raises some important questions for future investigation. Can online IRs produce emotional energy rivaling that generated in F2F contexts; if so, under what conditions? How effective is the translation of emotional energy across different IR media (e.g., online, F2F, broadcast)? What qualities of online interaction generate the greatest amount of emotional energy? Do online interactions simply serve to recharge and sustain the emotional energy produced in F2F interactions? Or, do online interactions engender a qualitatively distinct emotional energy that can be recycled across different media, and in F2F contexts?

While our discussion highlights the promise of extending IRT into online interaction, and points to areas for future investigation, the study of parasociality also has something to gain from our analysis. An IRT that incorporates the parasocial IR suggests that interactions across diverse technological media, as well as F2F, are creating a "convergence culture" (Jenkins 2008), where the dynamics of interaction of one media cannot be understood without examining others. What happens both on the screen and F2F are increasingly intertwined. It is our hope that a theory of IRs that incorporates TMI can help illuminate the parasocial dynamics of emotional energy in an increasingly mediated social world.

REFERENCES

Alperstein, N. (1991) Imaginary Social Relationships With Celebrities Appearing in Television Commercials. *Journal of Broadcasting and Electronic Media*, 35, 43–58.

Attwood, F. (2009) 'deepthroatfucker' and 'Discerning Adonis': Men and Cybersex. *International Journal of Cultural Studies*, 12(3), 279–294.

Auter, P. (1992) TV That Talks Back: An Experimental Validation of a Parasocial Interaction Scale. *Journal of Broadcasting and Electronic Media*, 36(2), 173–181.

Bainbridge, W. (2010) *Warcraft Civilization: Social Science in a Virtual World*. Cambridge: MIT Press.

Ballantine, P. and Martin, B. (2005) Forming Parasocial Relationships in Online Communities. *Advances in Consumer Research*, 32, 197–201.

Banos, R., Botella, C., Alcaniz, M., Liano, V., Guerrero, B. and Rey, B. (2004) Immersion and Emotion: Their Impact on the Sense of Presence. *CyberPsychology and Behavior*, 7(6), 734–741.

Bartle, R. (2004) *Designing Virtual Worlds*. Berkeley, CA: New Riders Publishing.
Bente, G., Rüggenberg, S., Krämer, N. and Eschenburg, F. (2008) Avatar-Mediated Networking: Increasing Social Presence and Interpersonal Trust in Net-Based Collaborations. *Human Communication Research*, 34(2), 287–318.
Ben-Ze'ev, A. (2004) *Love Online: Emotions on the Internet*. Cambridge: Cambridge University Press.
Boyns, D. (2010) The Savory Deviant Delight: A Study of Trust and Normative Order in an Online World. In D. Latusek and A. Gerbasi (Eds.), *Trust and Technology in a Ubiquitous Modern Environment: Theoretical and Methodological Perspectives*. Hershey, PA: IGI Global.
Boyns, D., Sosnovskya, E. and Forghani, S. (2009) MMORPG Worlds: On the Construction of Social Reality in World of Warcraft. In D. Heider (Ed.), *Living Virtually: Researching New Worlds*. New York: Peter Lang Publishing.
Brown, M. (1994) *Soap Opera and Women's Talk: The Pleasure of Resistance*. Thousand Oaks, CA: SAGE Publications.
Canary, D. and Spitzberg, B. (1993) Loneliness and Media Gratification. *Communication Research*, 20, 800–821.
Castronova, E. (2005) *Synthetic Worlds*. Chicago: University of Chicago Press.
Castronova, E. (2007) *Exodus to the Virtual World*. New York: Palgrave Macmillan.
Caughey, J. (1984) *Imaginary Social Worlds: A Cultural Approach*. Lincoln, NE: University of Nebraska Press.
Chayko, M. (2002) *Connecting: How We Form Social Bonds and Communities in the Internet Age*. Albany, NY: State University of New York Press.
Cohen, J. (1997) Parasocial Relations and Romantic Attraction: Gender and Dating Status Differences. *Journal of Broadcasting and Electronic Media*, 41, 516–529.
Cohen, J. (2004) Parasocial Breakup From Favorite Television Characters: The Role of Attachment Styles and Relationship Intensity. *Journal of Social and Personal Relationships*, 21, 187–202.
Coleman, G. (2010) The Hacker Conference: A Ritual Condensation and Celebration of a Lifeworld. *Anthropological Quarterly*, 83, 47–72.
Collins, R. (1981) On the Microfoundations of Macrosociology. *American Journal of Sociology*, 86(5), 984–1014.
Collins, R. (1993) Emotional Energy as the Common Denominator of Rational Action. *Rationality and Society*, 5(2), 203–230.
Collins, R. (2004) *Interaction Ritual Chains*. Princeton, NJ: Princeton University Press.
Collins, R. (2011) Interaction Rituals and the New Electronic Media. *The Sociological Eye*. Retrieved August 28, 2011 from http://sociological-eye.blogspot.com/2011/01/interaction-rituals-and-new-electronic.html
Derks, D., Fischer, A. and Bos, A. (2008) The Role of Emotion in Computer-Mediated Communication: A Review. *Computers in Human Behavior*, 24, 766–785.
Dibbell, J. (1998) *My Tiny Life: Crime and Passion in a Virtual World*. New York: Holt.
Durkheim, E. (1912/1995) *The Elementary Forms of Religious Life*. New York: The Free Press.
Eyal, K. and Cohen, J. (2006) When Good Friends Say Goodbye: A Parasocial Breakup Study. *Journal of Broadcasting and Electronic Media*, 50(3), 502–523.
Goffman, E. (1967) *Interaction Ritual: Essays on Face-to-Face Behavior*. New York: Anchor Books.

Hai-Jew, S. (2009) Exploring the Immersive Parasocial: Is it You or the Thought of You? *Journal of Online Learning and Teaching*, 5(3), 550–561.

Hanneman, R. and Collins, R. (1998) Modeling Interaction Ritual Theory of Solidarity. In P. Doreian and T. Fararo (Eds.), *The Problem of Solidarity: Theories and Models* (pp. 213–238). New York: Gordon and Breach.

Hardey, M. (2002) Life beyond the screen: embodiment and identity through the internet. *The Sociological Review*, 50(4), 570–585.

Hardey, M. (2008) The Formation of Social Rules for Digital Interactions. *Information, Community and Society*, 11(8), 1111–1131.

Harrington, C. and Bielby, D. (1995) *Soap Fans: Pursuing Pleasure and Making Meaning in Everyday Life*. Philadelphia: Temple University Press.

Henderson, S. and Gilding, M. (2004) "I've Never Clicked This Much With Anyone in My Life": Trust and Hyperpersonal Communication in Online Friendships. *New Media and Society*, 6(4), 487–506.

Horton, D. and Wohl, R. (1956) Mass Communication and Para-Social Interaction: Observations on Intimacy at a Distance. *Psychiatry*, 19, 215–229.

Huizinga, J. (1950) *Homo Ludens: A Study of the Play-Element in Culture*. Boston: The Beacon Press.

Jenkins, H. (2008) *Convergence Culture: Where Old and New Media Collide*. New York: NYU Press.

Jones, S. (1995) *Cybersociety: Computer-Mediated Communication and Community*. Thousand Oaks, CA: SAGE Publications.

Juul, J. (2005) *Half-Real: Video Games Between Real Rules and Fictional Worlds*. Cambridge: MIT Press.

Kassing, J. and Sanderson, J. (2009) "You're the Kind of Guy That We All Want For a Drinking Buddy": Expressions of Parasocial Interaction on Floydlandis.com. *Western Journal of Communication*, 73(2), 182–203.

Kendall, L. (1998) Meaning and Identity in 'Cyberspace': The Performance of Gender, Class and Race Online. *Symbolic Interaction*, 21(2), 129–153.

Kumar, N. and Benbasat, I. (2002) Para-Social Presence and Communication Capabilities of a Web Site: A Theoretical Perspective. *e-Service Journal*, 1(3), 5–24.

Leimeister, J.M., Schweizer, K., Leimeister, S. and Krcmar, H. (2008) Do Virtual Communities Matter for the Social Support of Patients?: Antecedents and Effects of Virtual Relationships in Online Communities. *Information Technology and People*, 21(4), 350–374.

Lee, K. (2004) Presence, Explicated. *Communication Theory*, 14, 27–50.

Lee, K. and Nass, C. (2005) Social-Psychological Origins of Feelings of Presence: Creating Social Presence With Machine-Generated Voices. *Media Psychology*, 7, 31–45.

Ling, R. (2008) *New Tech, New Ties: How Mobile Communication Is Reshaping Social Cohesion*. Cambridge: MIT Press.

Lombard, M. and Ditton, T. (1997) At the Heart of it All: The Concept of Presence. *Journal of Computer-Mediated Communication*, 3(2). Retrieved August 10, 2012 from http://www.ascusc.org/jcmc/vol3/issue2/lombard.html

Nardi, B. (2010) *My Life as a Night Elf Priest: An Anthropological Account of World of Warcraft*. Ann Arbor, MI: University of Michigan Press.

Nass, C. and Moon, Y. (2000) Machines and Mindlessness: Social Responses to Computers. *Journal of Social Issues*, 56(1), 81–103.

Nieckarz, P. (2005) Community in Cyber Space?: The Role of the Internet in Facilitating and Maintaining a Community of Live Music Collecting and Trading. *City and Community*, 4(4), 1–21.

Nowak, K. and Biocca, F. (2003) The Effect of the Agency and Anthropomorphism on Users' Sense of Telepresence, Co-Presence, and Social Presence in Virtual Environments. *Presence: Teleoperators and Virtual Environments*, 12, 2–35.

Raessens, J. (2005) Computer Games as Participatory Media Culture. In J. Raessens and J. Goldstein (Eds.), *Handbook of Computer Game Studies* (pp. 373–388). Cambridge: The MIT Press.

Reeves, B. and Nass, C. (1996) *The Media Equation: How People Treat Computers, Television, and New Media Like Real People and Places.* New York: Cambridge University Press/CSLI.

Rheingold, H. (1993) *The Virtual Community: Homesteading on the Virtual Frontier.* New York: Harper Perennial.

Riva, G., Mantovani, F., Capideville, C., Preziosa, A., Morganti, F., Villani, D., Gaggioli, A., Botella, C. and Alcaniz, M. (2007) Affective Interactions Using Virtual Reality: The Link Between Presence and Emotions. *Cyberpsychology and Behavior,* 10(1), 45–56.

Rubin, A. and Perse, E. (1987) Audience Activity and Soap Opera Involvement: A Uses and Effects Investigation. *Human Communication Research,* 14, 246–268.

Salen, K. and Zimmerman, E. (2003) *Rules of Play: Game Design Fundamentals.* Cambridge: The MIT Press.

Schiano, D.J., Nardi, B., Debeauvais, T., Ducheneaut, N. and Yee, N. (2011) *A New Look at World of Warcraft's Social Landscape.* Paper presented at the Foundations of Digital Games Conference, Bordeaux, France, June.

Tan, E. (2008) Entertainment is Emotion: The Functional Architecture of the Entertainment Experience. *Media Psychology,* 11(1), 28–51.

Taylor, T. (2006) *Play Between Worlds: Exploring Online Game Culture.* Cambridge: MIT Press.

Teoh, K. and Cyril, E. (2008) The Role of Presence and Para Social Presence on Trust in Online Virtual Electronic Commerce. *Journal of Applied Sciences,* 8, 2834–2842.

Turkle, S. (1995) *Life on the Screen: Identity in the Age of the Internet.* New York: Simon and Schuster.

Turkle, S. (2011) *Alone Together: Why We Expect More From Technology and Less From Each Other.* New York: Basic Books.

Turner, J. and Stets, J. (2005) *The Sociology of Emotions.* Cambridge: Cambridge University Press.

Yee, N. (2006) Motivations for Play in Online Games. *CyberPsychology and Behavior,* 9, 772–775.

3 Measuring Emotions in Individuals and Internet Communities

Dennis Küster and Arvid Kappas

INTRODUCTION

In recent years, researchers from fields as diverse as psychology, computer sciences, and communication sciences have increasingly turned to the Internet as a source of data about emotions. For psychology, studying emotions on the Internet is a new and exciting challenge because this medium offers vast amounts of data and greater ecological validity than most laboratory experiments. However, "online emotions" are typically observed only indirectly in some form of textual communication recorded on the Internet, rather than more immediately in the emotional or physiological responses of the human participants of a conversation. While this may be less of a concern from a pure communications perspective, the relationship between the textual content of online communication and the emotional states of people turns into a key issue for the study of psychological processes associated with such emotions.

In this chapter, we view online emotions from the vantage point of psychology, and outline how emotions shared on the Internet can be measured in three principal ways that are not mutually exclusive but have individual strengths and weaknesses: first, researchers can study large amounts of emotional content on the Internet. Second, they can ask individuals about their emotional experience online. Third, they can record bodily responses to measure emotions unobtrusively. We will discuss how these three methods can complement each other, with a particular focus on the potential of text-based analyses on the Internet. At the same time, we will emphasize some of the critical limitations. While we focus on how this data is related to human emotional experience, processing, and behavior, we believe that a broader perspective and awareness of methodological possibilities can be of value to the study of online emotions in general. Toward this goal, we conclude the chapter with a brief discussion of a number of new applications and developments that promise to deliver behavioral data relating to emotions on the Internet beyond the text.

ONLINE EMOTIONS

Online emotions refer to a field of study that is primarily concerned with the analysis of large amounts of data that are already present somewhere on the Internet[1]. Typically, such data exists in the form of texts that have originated in some form of online interaction, be it in a chat, a blog, a forum, Twitter, Facebook, or any other of the multitude of online communication instruments that have emerged on the Internet. It is this kind of data that characterizes natural emotional communication behavior as it has emerged on the Internet, and that promises insights into dynamics and phenomena unique to this medium that might not be found elsewhere. An advantage of this kind of approach is that it clearly targets online phenomena. However, a drawback is that the link to "emotions" is more tenuous. In part, this has historical reasons, and in part this has to do with the properties of one of the most popular methods: automated text analysis. While our focus in this chapter is on the potential of such automated analyses, we will discuss how methods that have originally been developed for an "offline" context may still be able to make important contributions to our understanding of online emotions.

AUTOMATED TEXT ANALYSES

Current trends in measuring online emotions have a strong focus on automated collection and analysis of text-based data sets at a previously almost unimaginable scale. Where 10 years ago (e.g., Tidwell and Walther 2002), researchers would investigate the content of individual emails collected in the laboratory, today the Internet itself has become the laboratory (Skitka and Sargis 2006; Chesney et al. 2009). Data on the Internet can be investigated on a large scale with the help of publicly-available algorithms and software like the LIWC (Linguistic Inquiry and Word Count, Pennebaker et al. 2007) or SentiStrength (Thelwall et al. 2010, downloadable from http://sentistrength.wlv.ac.uk). The success of this approach has been demonstrated, for example, by studies that have measured mood on a national level (Mislove et al. 2010), used Twitter-mood to predict the stock market (Bollen et al. 2011), or extracted and compared diurnal and seasonal mood rhythms across cultures (Golder and Macy 2011).

A particular benefit of automated emotion measures is the possibility to relate recent or even real-time data of millions of online statements to other psychologically interesting data, including information about the social networks themselves (e.g., Kwak et al. 2010). In the case of complete data sets of a given domain, careful data analysis may furthermore effectively offset some of the traditional disadvantages associated with the inherent lack of experimental control groups of data collected "in the field." For example, Danescu-Niculescu-Mizil et al. (2011) took a data set of 1.3 million Twitter conversations, out of which they extracted "just" 215,000 conversations between 2,200 pairs of users that satisfied certain criteria of reciprocity. Such mutual reciprocity may not be entirely typical for Twitter, but there may still be a sufficient number of instances to allow powerful statistical tests.

In light of these advantages, automated analyses may appear to be the method of choice for a large number of research questions. However, they also face a number of limitations. For example, the usage distribution and the probability of someone to be included in a sample may be extremely skewed. This can lead to a sampling bias where a small number of subjects have a disproportionately large effect on any results that an online researcher would like to interpret as characteristic for an entire online community or network. It has been shown that the degree to which individuals in a given community participate in, for example, a forum (Si and Liu 2010), blogs (Mitrović et al. 2011), or Twitter is typically characterized by a power law (Asur and Huberman 2010) rather than a normal distribution. In other words, some people contribute enormous amounts of data, others very little.

This is in line with the more general assertion that a few early emotional statements by influential nodes may have a great influence on overall "mood"—whereas other nodes may have comparatively little impact (Barabási and Albert 1999). In this light, it may not be entirely surprising if a single tweet by Justin Bieber can motivate 40,000 people to follow a new node within the network—within only 15 minutes (Schroeder 2012). Such a power-law type distribution means that certain individuals may have a greatly exaggerated probability of being included in the sample compared to other individuals who have been either less vocal, or who perhaps have simply received less attention from the community. For example, given that it has been found that personality significantly influences Facebook usage (Moore and McElroy 2011), the frequency with which data from people with a certain type of personality is sampled may be much higher than that of others who post less often. In general, this kind of issue does not appear in the laboratory where each participant typically produces a very similar amount of data.

From a slightly different perspective, this issue may also be regarded as an interesting property of online emotions that may lead to different social dynamics than what is normally observed for "offline emotions." For example, it has been demonstrated that emotions expressed in posts in an online network like BBC forums do not occur randomly and independently of one another—rather, emotions of the same valence tend to occur together, leading to non-random chains (Chmiel et al. 2011). In consequence, emotions in online discussions may spread more rapidly, and perhaps differently, than emotions in offline discussions, where the boundaries between individual and mass communication are more clearly drawn.

How Automated Text Analyses Work

Automated analyses are attractive because they can measure very large amounts of data in comparatively little time. Researchers can easily use "off the shelf" software that is based on extensive emotion dictionaries, and begin their analysis almost immediately. An example of such standalone software is the LIWC, which can be used to classify any sort of online text for emotional content and relative frequency of words. Other comparable software, e.g., SentiStrength, can be used to efficiently classify emotional content, even of short informal texts of texts (Thelwall et al. 2010).

In research that focuses more specifically on the issue of "sentiment" detection in online texts (see e.g., Paltoglou and Thelwall 2011), however, the analysis rarely stops at this point. For example, a certain degree of context can be accounted for via machine-learning by estimating the "contextual polarity" of the phrase in which a word is contained (Wilson et al. 2009). Here, the aim is to automatically detect when words that have a "prior" polarity (positive, neutral, or negative) in isolation change their polarity (i.e., valence) within a particular phrase. For example, negative or positive words that are used as part of an expression may be neutral rather than express an emotion—or algorithms can classify features of preceding words (adjectives, adverbs, intensifiers, etc.; for further examples, see Wilson et al. 2009). In combination with validation against human annotations, this may be a promising approach to make machine-learning more context sensitive (see Wilson et al. 2009). Furthermore, this kind of machine-learning can include optimization by "supervision" (Sebastiani 2002) in the form of manual annotation of a training subset of the texts to be analyzed (e.g., Chmiel et al. 2011). In brief, such training subsets are used to optimize, validate, and adjust the classification algorithms to the specific type of texts and contexts in question (Sebastiani 2002).

While it is possible to further analyze, visualize, and contextualize textual data via other means like network analyses, questions about individual or collective online emotions eventually lead to issues that are more psychological in nature. In complex network analyses, user communities can be identified and mapped onto network structures that mimic online communication across time, e.g., "bipartite networks," where nodes are divided into two sets that can accurately represent incoming and outgoing links between users (Mitrović et al. 2011). Studies like these can show, for example, that a small fraction of very active users may be able to critically tune emotions within the larger network. Yet it remains a psychological question to determine why, precisely, those "nodes" might behave differently in the first place.

When we look at the high level of agreement with a human rater shown for some of the automated text analysis tools (e.g., Thelwall et al. 2011), there initially appears to be little need for further improvement. An issue, however, is the question of what precisely is classified as emotional. If, for

52   *Dennis Küster and Arvid Kappas*

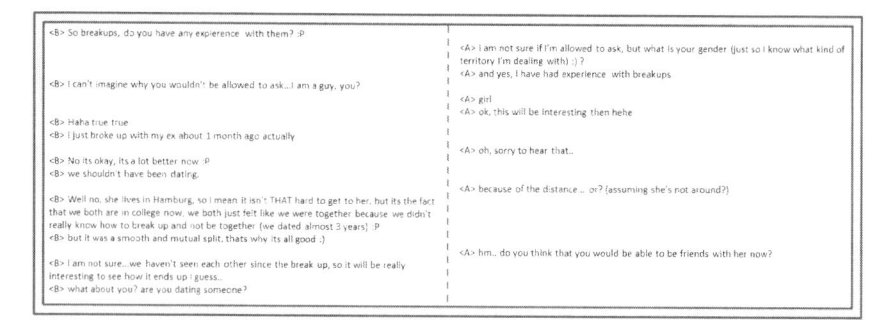

```
<B> So breakups, do you have any experience with them? :P
                                                                <A> I am not sure if I'm allowed to ask, but what is your gender (just so I know what kind of
                                                                territory I'm dealing with) :) ?
<B> I can't imagine why you wouldn't be allowed to ask...I am a guy, you?
                                                                <A> and yes, I have had experience with breakups
                                                                <A> girl
<B> Haha true true                                              <A> ok, this will be interesting then hehe
<B> I just broke up with my ex about 1 month ago actually
                                                                <A> oh, sorry to hear that..
<B> No its okay, its a lot better now :P
<B> we shouldn't have been dating.
                                                                <A> because of the distance .. or? (assuming she's not around?)
<B> Well no, she lives in Hamburg, so I mean it isn't THAT hard to get to her, but its the fact
that we both are in college now, we both just felt like we were together because we didn't
really know how to break up and not be together (we dated almost 3 years) :P
<B> but it was a smooth and mutual split, thats why its all good :)
                                                                <A> hm.. do you think that you would be able to be friends with her now?
<B> I am not sure...we haven't seen each other since the break up, so it will be really
interesting to see how it ends up I guess..
<B> what about you? are you dating someone?
```

Figure 3.1   Example of a dyad (here a male and a female) discussing the topic of relationship "breakup" in an online chat in the laboratory.

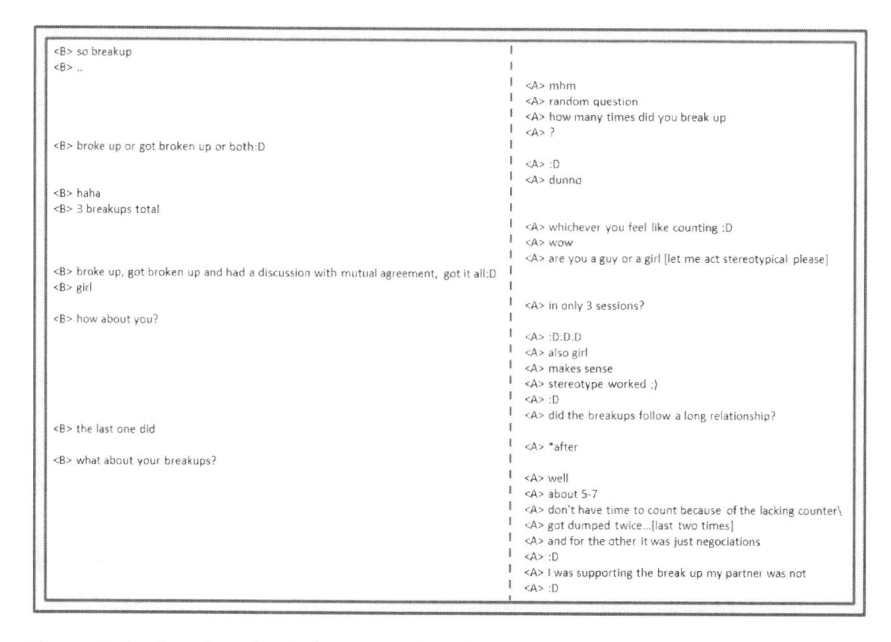

```
<B> so breakup
<B> ..
                                                                <A> mhm
                                                                <A> random question
                                                                <A> how many times did you break up
                                                                <A> ?
<B> broke up or got broken up or both:D
                                                                <A> :D
                                                                <A> dunno
<B> haha
<B> 3 breakups total
                                                                <A> whichever you feel like counting :D
                                                                <A> wow
                                                                <A> are you a guy or a girl [let me act stereotypical please]
<B> broke up, got broken up and had a discussion with mutual agreement, got it all:D
<B> girl
                                                                <A> in only 3 sessions?
<B> how about you?
                                                                <A> :D:D:D
                                                                <A> also girl
                                                                <A> makes sense
                                                                <A> stereotype worked ;)
                                                                <A> :D
                                                                <A> did the breakups follow a long relationship?
<B> the last one did
                                                                <A> *after
<B> what about your breakups?
                                                                <A> well
                                                                <A> about 5-7
                                                                <A> don't have time to count because of the lacking counter\
                                                                <A> got dumped twice...[last two times]
                                                                <A> and for the other it was just negotiations
                                                                <A> :D
                                                                <A> I was supporting the break up my partner was not
                                                                <A> :D
```

Figure 3.2   Another dyad (here two females) discussing the topic of relationship "breakup."

example, the persons manually coding the training data do not have sufficient information about the context of a conversation, then the human classification may be a rather poor reflection of the emotions experienced by the participants of that conversation. The text classification algorithm might still produce a good match with this human "gold standard"—but the apparent emotions in the text would still be misunderstood.

At this point, we might take a look at two short cases from our own laboratory where we have asked dyads of participants to discuss emotional topics via a chat interface. In both of these conversations (see Figures 3.1

and 3.2), participants were discussing the topic of relationship "breakup." A human observer who sees both discussions might quickly infer, for example, that not all of the humor and emoticons used here necessarily reflect strong positive emotions that are experienced at that moment by a participant. Rather, at least some of these might be active efforts by one of the participants to regulate potentially negative emotions that might otherwise be associated with talking about such unpleasant and intimate experiences.

This, of course, does not mean that the algorithm would necessarily be wrong—it might still be able to distinguish more or less pleasant exchanges about the same topic if given sufficient data. For example, *SentiStrength* uses tokens with values that can be modified through the analysis of additional documents. We have also seen how simple features at the level of the phrase may help to determine the polarity of the context (Wilson et al. 2009), and this could help to further reduce ambiguity. Alternatively, we might employ another software like the LIWC to derive an automatic count of various psychologically interesting classes of words that can be meaningfully interpreted if we have a large enough sample. Depending on the precise software we use, different characteristics of the text can be brought into focus in a quantitative way. However, it also appears clear that exchanges such as these are not easy to evaluate with certainty even for human raters, perhaps not even for the other partner in each dyad who will know somewhat more about the general context from the previous discussions, from living on the same campus, or by having the same gender. It is perhaps no surprise, therefore, that in both of these examples, participants eventually ask about the other participant's gender—and that emoticons and humor are used with such frequency (see also Tidwell and Walther 2002).

Limits of Automated Text Analyses and the Issue of Context

As we have just seen, the unique environment of online communication may shape online emotions in subtle ways that can be an interesting object of study in their own right. However, the problem is that, eventually, researchers almost always want to draw conclusions about the real world—about the emotions, experiences, behaviors, and decisions of the humans involved in online communication. From a methodological perspective, this is where issues of context and meaning of text analyses become much more difficult.

As it turns out, even in the rare case where multiple measures of emotion can be taken directly from participants in the laboratory, there is surprisingly little agreement between them. Thus, as indicated in a recent review (Mauss and Robinson 2009), cohesion between different measures of emotion can, at best, be expected to be moderate. This even applies for the case where facial muscle activity is measured via electrodes and related to self-report (see, e.g., Larson et al. 2003), or where facial actions are coded reliably from a recording (Krumhuber and Manstead 2009; Reisenzein 2000).

How does this relate to "context" in the wider sense? A simple example of the importance of context for measuring emotion is the emotional classification of a facial expression found in a picture. In psychological studies (e.g., Barrett et al. 2011), individual expressions in photos can easily be separated from their context so that even human raters can no longer accurately judge the emotional state of the person(s) depicted. Likewise for online texts: Take e.g., a simple smiley emoticon. Clearly, a smiley can be highly ambiguous when the question we ask is what emotional state the author is experiencing at that moment. When participants interact in an online chat in the laboratory, for example, the smiley may appear when they are greeting one another, when someone has just made a joke, or when a participant has just deleted a half-finished statement, to mention just a few possibilities. Smiley emoticons are even frequently found in conversational closings, when people are saying goodbye online (Pojanapunya and Jaroenkitbowron 2011). Here, what one might intuitively perceive as the best online equivalent of a smile, may perhaps be just that. The problem is: neither the online *nor* the offline version seems to distinguish very well between such fundamentally different psychological entities such as happiness, politeness, or shame. It is important to keep in mind that people do not write text online to simply give a read-out of how they feel. Instead, as is the case in offline situations, messages are *actions* in a conversational process (e.g., Austin 1975). Senders want to achieve a variety of goals and they do this by choosing what they communicate and how—to a certain degree. This is where it becomes interesting. There is reason to believe that communicators are not aware of certain structural aspects of their language, such as the frequency of certain word types, or choice of particular words, that nevertheless might have a particular impact in the receiver. In any case, researchers typically do not have access to the communication intent.

In some instances, this may not be very relevant. If, for example, the aim of a study is an investigation of daily mood rhythms expressed across entire cultures (Golder and Macy 2011), researchers may not need to distinguish politeness or happiness because for a broad definition of "mood" this may not matter and expression can be seen in the widest sense as a feature of the message and not of the sender. Furthermore, unless politeness has a different rhythm than happiness in some cultures but not others—a distinction is not necessary because the main finding would remain the same. In many other cases, however, we do want to be able to make these kinds of distinctions. And, unless we are dealing with data from our own culture, this can be a difficult task even for humans.

SUBJECTIVE AND PHYSIOLOGICAL MEASURES OF ONLINE EMOTIONS

We have already discussed the issue of context as an important limiting factor for the potential of automated text analysis tools to correctly interpret and classify online emotions. Furthermore, we have seen that a basic component needed for the creation or improvement of any new automatic classifier is the "training" of the classifier by data that has been annotated by humans. This is the point where large-scale quantitative and probabilistic algorithms generally connect back to small-scale data sampling where human raters are required to read, understand, and annotate the data. In a sense, this is also what links online to "offline" emotions for automated text analyses tools. Without this link, it would appear difficult to justify that the object of study in online emotions, as extracted by automated tools, indeed has much to do with "emotion" at all. This leads us to the question of what else subjective response measures, i.e., human evaluations might be able to contribute to establish a "ground-truth" that is not based on rater agreement.

Subjective Response Measurement

When we talk about subjective evaluations of emotional online communication, one of the first things to note is that reading is usually not the only activity that online communicators engage in. To read is what the human annotators of training data do, and it is what the text analyses tools do. However, at least some of the participants of online communication also engage actively in emotional discussions—and their emotional experience might clearly be rather different from a passive reader who primarily tries to be a good annotator. In other words, when we discuss online emotions, we should not forget that what we are dealing with follows the basic structure of a conversational model. In consequence, there are primarily three components that can be measured: the emotional content of online materials as such (the message), the emotions of authors (senders), and the emotional states of readers (receivers). Even if we decide to focus on just one of the components, e.g., the message, our understanding will be rather limited if we have no idea about the other two.

In consequence, it may be useful to ask a larger number of participants directly how they felt when reading or writing emotional exchanges (e.g., Kappas et al. 2011). However, this too has limitations. For example, certain aspects of emotional experience may be difficult to detect, verbalize, or quantify for most people without specific training (Sze et al. 2010). Furthermore, there is evidence showing that people may generally be better at reporting current emotions rather than emotional experiences that are distanced in time (Robinson and Clore 2002). If we relate this again to automated text analyses, this certainly does not invalidate self-report or human ratings as a whole (see also Mauss and Robinson 2009). However, the human gold standard may only be valid under certain conditions.

Nevertheless, a more explicit consideration of the potential roles of subjective evaluations might help to improve automated analyses as well as to increase our understanding of some of the differences that can be observed across multiple online communication environments. This could include collecting data on how individuals in online communities subjectively feel when, e.g., a text-based algorithm suggests a certain result. In a relatively minimal version, this could mean that at least some of the human raters used to train a given classifier should themselves be part of the community in question (see Thelwall et al. 2010). For example, positive sentiments have been shown to be characteristic for MySpace (Thelwall et al. 2010), whereas chains of negative comments have been found to be important on BBC forums (Chmiel et al. 2011). A human rater who has been heavily participating in MySpace may perceive a statement in the typical tone of a political discussion forum very differently than a rater who has substantial experience on such a form.

Specialized tools and methods can also be developed to better suit different online contexts, including more qualitative approaches. For example, communication and sharing of emotions via Facebook may contain much more visual information that is of personal relevance to the users than a news forum. While we have structured this chapter along the principal sources of data rather than the quantitative vs. qualitative dichotomy in the analysis of texts that can be found across disciplines, this is also a point where quantitative research might greatly benefit from insights derived on a qualitative basis.

Physiological Measures

While an analysis of the texts that people write combined with subjective response measurement of emotional experiences may be a cornerstone of online emotion research, what is missing in this picture are the "bodily emotions" of individuals and communities on the Internet. Why is this important?

Unfortunately, a physiological "gold-standard" still seems to be some time away at best—at the worst, there will never be a gold-standard due to physiological idiosyncrasies. It is an ongoing debate how much information physiological changes in the periphery or in the brain itself can provide regarding specific emotional states. According to the current state of the art, high correlations with subjective measures are typically rare, even under highly controlled conditions in the laboratory (e.g., Mauss and Robinson 2009). However, physiological measures play a critical role in furthering our conceptual understanding of emotional processes offline as well as online (see also Calvo and D'Mello 2010; Kappas 2010).

It is important to remember that emotions are primarily self-regulating processes—i.e., their function is to regulate and guide behavior of the self and others (Kappas 2011). For example when people are merely reading emotional exchanges between other people (e.g., Theunis et al. 2010), this can be investigated in the smiles, frowns, and heart rate changes that are consistently elicited as a part of the emotional response and regulation processes online—even in observers of emotional exchanges. This relates also to processes such as empathy or counterempathy with others and this renders the interpretation of changes in physiological activation ever-more complex.

We argue that physiological measures may play a critical role in clarifying the conceptual underpinnings of online emotions. Apart from their importance for emotion theory, however, physiological measures have also been in increasing demand for their use in practical applications. This leads to our concluding discussion regarding new directions in online emotion research.

CONCLUSIONS AND NEW DIRECTIONS IN ONLINE MEASUREMENT

As we have argued in this chapter, online emotion measures based on text appear to hold great promise for the study of online emotions. This is, to a large extent, due to the possibility of collecting huge data sets of primarily textual exchanges that can be automatically classified and dissected for their emotional content. However, we have also seen that the limitations that algorithms as well as humans face when dealing with partially unknown or even unknowable social contexts can lead to systematic biases. We have argued that one important building block aimed at dealing with this issue must be substantive basic research that takes a clear account of existing limitations and exposes at least some of the presently still insufficiently studied underpinnings of the processes in question. Without additional data, theory and models that can tease apart and distinguish emotional behavior guided, e.g., by politeness, pleasantness, or social goals, an understanding of online "emotions" that is still partially built on limited and sometimes untested foundations will itself remain shaky (see also Kappas 2010). At worst, an inflated impression of accuracy can even be misleading—e.g., when a seemingly polite level of conversation within a community is misunderstood as unemotional, or vice versa.

On the practical side, however, there have also been recent innovations that afford new possibilities for applied and theoretical research on a scale comparable to that of large-scale text-based emotion analyses. For example, a new generation of algorithms and tools is beginning to try to tap into real-time video information that can be recorded by simple and ubiquitously available webcams. Such algorithms can already be used to provide estimates of autonomic measures like heart rate and heart rate variability (e.g., Poh et al. 2011), or facial activity (Picard 2010; Bartlett and Whitehill 2011; see also Calvo and D'Mello 2010). While this is unlikely to quickly solve some of the conceptual problems as such due to low cohesion between even highly reliable measures of facial activity and subjective report (see e.g., Reisenzein et al. 2006), it seems clear that such tools may have the potential to revolutionize the future of online emotion measurement using multimodal approaches. Devices that tap into gestures and body movement for interfacing with computers are likely to become increasingly available after the success with which Microsoft Kinect was adopted by many researchers for creative interface design. Clearly, there is much potential for the analysis of affective states in modulating gestures and postures.

Once basic research in the laboratory has disentangled some of the factors underlying real-time emotional behavior online, such tools might be used to test resulting predictions in the field—potentially on a large scale. At present, it still remains to be seen to what extent new applications (e.g., also Apple's natural language interface for iPhones, Siri) involving automated text or video analyses will be accepted and used by the mass market. However, it appears likely that the outlook on measures of emotional or face-to-face communication over the Internet (see also Kappas and Krämer 2011) will involve an increasing amount and sophistication of algorithmically-supported analyses. From the research perspective on methods of measuring emotions on the Internet, chances appear to be that the Internet might indeed soon expand and evolve to the point where emotion measurement at multiple levels (e.g., textual, subjective, and physiological) can be obtained with all the advantages (and drawbacks) of large-scale research that we have discussed in this chapter.

AUTHOR NOTE

Research from our laboratory that is reported in this manuscript was supported by the CYBEREMOTIONS: EU FP7-funded project, Theme 3: "Science of complex systems for socially intelligent ICT," contract 231323.

NOTES

1. The Internet can also serve as a medium for experimental- or questionnaire-type studies where researchers are in direct contact with the users (e.g., Skitka and Sargis 2005; Hewson 2007; Reips 2007), but this is not strictly about emotion communication on the Internet.

REFERENCES

Asur, S. and Huberman, B.A. (2010) *Predicting the Future With Social Media.* arXiv:1003.5699v1
Austin, J.L. (1975) *How to do Things With Words* (2nd ed.). Oxford: Oxford University Press.
Barabási, A.-L. and Albert, R. (1999) Emergence of Scaling in Random Networks. *Science,* 286, 509–512. doi:10.1126/science.286.5439.509
Barrett, L.F., Mesquita, B. and Gendron, M. (2011) Context in Emotion Perception. *Current Directions in Psychological Science,* 20, 286–290. doi:10.1177/0963721411422522

Bartlett, M.S. and Whitehill, J. (2011) Automated Facial Expression Measurement: Recent Applications to Basic Research in Human Behavior, Learning, and Education. In A.J. Calder, G. Rhodes. M.H. Johnson, and J.V. Haxby (Eds.), *Oxford Handbook of Face Perception* (pp. 489–513). Oxford: Oxford University Press.

Bollen, J., Mao, H. and Zeng, X. (2011) Twitter Mood Predict the Stock Market. *Journal of Computational Science*, 2, 1–8. doi:10.1016/j.jocs.2010.12.007

Calvo, R.A. and D' Mello, S.K. (2010) Affect Detection: An Interdisciplinary Review of Models, Methods, and Their Applications. *IEEE Transactions on Affective Computing*, 1, 18–37.

Chmiel, A., Sienkiewicz, J., Thelwall, M., Paltoglou, G., Buckley, K., Kappas, A. and Hołyst, J. A. (2011) Collective Emotions Online and Their Influence on Community Life. *PLoS ONE*, 6, e22207. doi:10.1371/journal.pone.0022207

Danescu-Niculescu-Mizil, C., Gamon, M. and Dumais, S. (2011) Mark My Words! Linguistic Style Accommodation in Social Media. arXiv:1105.0673v1

Golder, S.A. and Macy, M.W. (2011) Diurnal and Seasonal Mood Vary With Work, Sleep, and Daylength Across Diverse Cultures. *Science*, 333, 1878. doi: 10.1126/science.1202775

Hewson, C. (2007) Gathering Data on the Internet: Qualitative Approaches and Possibilities For Mixed Methods and Research. In A.N. Joinson, K.Y.A. McKenna, T. Postmes, and U.-D. Reips (Eds.), *The Oxford Handbook of Internet Psychology* (pp. 405–408). Oxford: Oxford University Press.

Kappas, A. (2010) Smile When You Read This, Whether You Like It or Not: Conceptual Challenges to Affect Detection. *Social Psychology*, 1, 38–41. doi:10.1109/T-AFFC.2010.6

Kappas, A. (2011) To Our Emotions, With Love: How Affective Should Affective Computing Be? *Affective Computing and Intelligent Interaction 4th International Conference 2011 Proceedings, Part I, Lecture Notes in Computer Science*, 6874, 1. doi:10.1007/978-3-642-24600-5_1

Kappas, A. and Krämer, N.C. (Eds.). (2011) *Face-to-Face Communication Over the Internet: Emotions in a Web of Culture, Language and Technology*. Cambridge: Cambridge University Press.

Kappas, A., Tsankova, E., Theunis, M. and Küster, D. (2011, September) *Cyber-Emotions: Subjective and Physiological Responses Elicited by Contributing to Online Discussion Forums*. Poster presented at the 51st Annual Meeting of the Society For Psychophysiological Research, Boston, Massachusetts.

Krumhuber, E.G. and Manstead, A.S.R. (2009) Can Duchenne Smiles be Feigned? New Evidence on Felt and False Smiles. *Emotion*, 9, 807–820. doi:10.1037/a0017844

Kwak, H., Lee, C., Park, H. and Moon, S. (2010) What is Twitter, a Social Network or a News Media? WWW '10 Proceedings of the 19th international conference on World Wide Web ACM, New York. doi: 10.1145/1772690.1772751

Larsen, J.T., Norris, C.J. and Cacioppo, J.T. (2003) Effects of Positive and Negative Affect on Electromyographic Activity Over Zygomaticus Major and Corrugator Supercilii. *Psychophysiology*, 40, 776–785. doi: 10.1111/1469-8986.00078

Mauss, I.B. and Robinson, M.D. (2009) Measures of Emotion: A Review. *Cognition and Emotion*, 23, 209–237. doi:10.1080/02699930802204677

Mislove, A., Lehmann, S., Ahn, Y., Onnela, J. and Rosenquist, J.N. (2010) Pulse of the Nation: U.S. Mood Throughout the Day Inferred From Twitter. doi: 10.1126/science.1202775

Mitrović, M., Paltoglou, G. and Tadić, B. (2011) Quantitative Analysis of Bloggers' Collective Behavior Powered by Emotions. *Journal of Statistical Mechanics: Theory and Experiment*, P02005. doi:10.1088/1742-5468/2011/02/P02005

Moore, K. and McElroy, J.C. (2011) The Influence of Personality on Facebook Usage, Wall Postings, and Regret. *Computers in Human Behavior*, 28, 267–274. Elsevier Ltd. doi:10.1016/j.chb.2011.09.009

Paltoglou G. and Thelwall, M. (2011) Twitter, MySpace, Digg: Unsupervised Sentiment Analysis in Social Media. *ACM Transactions on Intelligent Systems and Technology, Special Issue on Search and Mining User Generated Contents*.

Pennebaker, J.W., Booth, R.J. and Francis, M.E. (2007) Linguistic Inquiry and Word Count: LIWC 2007. Austin, TX: LIWC (www.liwc.net).

Picard, R.W. (2010) Emotion Research By the People, For the People. *Emotion Review*, 2, 250–254. doi: 10.1177/1754073910364256

Poh, M.-Z., McDuff, D.J. and Picard, R.W. (2011) Advancements in Noncontact, Multiparameter Physiological Measurements Using a Webcam. *IEEE Transactions on Bio-Medical Engineering*, 58, 7–11. doi:10.1109/TBME.2010.2086456

Pojanapunya, P. and Jaroenkitboworn, K. (2011) How to Say "Good-Bye" in Second Life. *Journal of Pragmatics*, 43, 3591–3602. doi:10.1016/j.pragma.2011.08.010

Reips, U.-D. (2007) The Methodology of Internet-Based Experiments. In A.N. Joinson, K.Y.A. McKenna, T. Postmes, and U.-D. Reips (Eds.), *The Oxford Handbook of Internet Psychology* (pp. 373–390). Oxford: Oxford University Press.

Reisenzein, R. (2000) Exploring the Strength of Association Between the Components of Emotion Syndromes: The Case of Surprise. *Cognition and Emotion*, 14, 1–38. doi:10.1080/026999300378978

Reisenzein, R., Bördgen, S., Holtbernd, T. and Matz, D. (2006) Evidence For Strong Dissociation Between Emotion and Facial Displays: The Case of Surprise. *Journal of Personality and Social Psychology*, 91, 295–315. doi: 10.1037/0022-3514.91.2.295

Robinson, M.D. and Clore, G.L. (2002) Episodic and Semantic Knowledge in Emotional Self-Report: Evidence For Two Judgment Processes. *Journal of Personality and Social Psychology*, 83, 198–215. doi:10.1037//0022–3514.83.1.198

Schroeder, S. (2012) With Help From Bieber, Matt Lauer Gets 40,000 Twitter Followers. *Mashable Social Media*. Retrieved June 15, 2012, from http://mashable.com/2012/06/15/matt-lauer-gets-40000-twitter-followers-in-minutes-thanks-to-justin-bieber/

Sebastiani, F. (2002) Machine Learning in Automated Text Categorization. *ACM Computing Surveys*, 34, 1–47. doi:10.1145/505282.505283

Si, X.-M. and Liu, Y. (2010) Power-Law Distribution of Human Behaviors in Internet Forums. *2010 International Symposium on Intelligence Information Processing and Trusted Computing*, 286–289. IEEE. doi:10.1109/IPTC.2010.79

Skitka, L.J. and Sargis, E.G. (2005) Social Psychological Research and the Internet: The Promise and Peril of a New Methodological Frontier. In Y. Amichai-Hamburger (Ed.), *The Social Net: The Social Psychology of the Internet* (pp. 1–29). Oxford: Oxford University Press.

Skitka, L.J. and Sargis, E.G. (2006) The Internet as Psychological Laboratory. *Annual Review of Psychology*, 57, 529–555.

Sze, J.A., Gyurak, A., Yuan, J.W. and Levenson, R.W. (2010) Coherence Between Emotional Experience and Physiology: Does Body Awareness Training Have an Impact? *Emotion*, 10, 803–814. doi:10.1037/a0020146

Thelwall, M., Buckley, K. and Paltoglou, G. (2011) Sentiment Strength Detection For the Social Web. *Journal of the American Society For Information Science and Technology*. doi: 10.1002/asi.21662

Thelwall, M., Buckley, K., Paltoglou, G., Cai, D. and Kappas, A. (2011) Sentiment Strength Detection in Short Informal Text. *Journal of the American Society For Information Science and Technology*, 61, 2544–2558. doi: 10.1002/asi.21416

Thelwall, M., Wilkinson, D. and Uppal, S. (2010) Data Mining Emotion in Social Network Communication: Gender Differences in MySpace. *Journal of the American Society For Information Science and Technology*, 21, 190–199. doi:10.1002/asi.21180

Theunis, M., Küster, D., Tsankova, E., & Kappas, A. (2010) *CyberEmotions: Online discussion forums elicit subjective emotional response*. Poster accepted for presentation at the 3rd European Conference on Emotion, organized by the Consortium of European Research on Emotion, Villeneuve d'Ascq, France.

Tidwell, L.C. and Walther, J.B. (2002) Computer-Mediated Communication Effects on Disclosure, Impressions, and Interpersonal Evaluations: Getting to Know One Another a Bit At a Time. *Human Communication Research*, 28, 317–348. doi: 10.1111/j.1468-2958.2002.tb00811.x

Wilson, T., Wiebe, J. and Hoffmann, P. (2009) Recognizing Contextual Polarity: An Exploration of Features For Phrase-Level Sentiment Analysis. *Computational Linguistics*, 35, 399–433. doi:10.1162/coli.08-012-R1-06-90

Part II
Emotions Display

4 Grief 2.0
Exploring Virtual Cemeteries
Nina R. Jakoby and Simone Reiser

"Grief is the sole comfort of the bereaved"
—Robert Hamerling

INTRODUCTION

As long as there are social and intimate relationships, friendship and love, there is grief. It is the price we pay for love, the "cost of commitment" (Parkes 1972) and therefore an elementary form of human experience (Archer 1999; Cochran and Claspell 1987; Walter 2000). Grief is integral to life and not a condition to be treated (see Valentine 2008: 3). From a sociological point of view, grief is seen as a social emotion and interpersonal process because it emerges from relationships, attachments, expectations, and obligations (see Charmaz and Milligan 2006: 525). The sociological model of grief focuses on *grief as an emotion* in its non-pathological form (Jakoby 2012). It is a prototype of "normal sadness" and a facet of human experience (Horwitz and Wakefield 2007).

Individuality, flexibility, and diversity are general features of contemporary society (Winkel 2001) that also form the complexity of postmodern bereavement. Moreover, influences of demographic transitions, social and geographical mobility, family fragmentation, secularization or deritualization shape the experience of grief (e.g., Walter 1996). Secularization and technological progress as well as the individualization and emotionalization of death are developments that have also shaped the history of the cemetery (for a more detailed account, see Ariès 2009). These processes are closely connected with the survivors' wish to preserve the memories of their loved ones and mourn the loss. Memorials express how society and the individual member relate to death in general and the deceased in particular (Geser 1998c). Media support remembrance and preservation and therefore play an essential role in expressing grief. The Internet, as a medium of recollection and a driver of social and cultural change, influences burial customs and offers virtual cemeteries as a new platform for languaging the emotion of grief. In the online world, we can observe sociological processes and theories in an ideal-type manner, which has led scholars to call the Internet a "methodological laboratory" (Geser 1998a, 2002). Hence, it offers ideal conditions for reviewing theories in the sociology of emotions, especially symbolic interactionism and the assumptions it makes with regard to grief and grieving. The strength of symbolic interactionism lies in creating awareness for the numerous rules governing the emotion of grief, in particular by distinguishing between socially legitimate and illegitimate losses and identifying the specific rules for "proper" grief, which derive from the psychological model of "normal grief". Can we identify these social rules of grief, and do we find a new representation of grief on the Internet? Based on content analyses of two German-speaking web memorials, we will investigate how the absence of physical parameters in the online world, such as space, time, and matter, affect how grief is expressed.

GRIEF AND SYMBOLIC INTERACTIONISM

Explicit theorizing about grief within the sociology of emotions is found only in the theory of symbolic interactionism (Charmaz 1980). This approach covers a wide range of ideas, but there are two major contributions to the understanding of grief that are discussed in this section. First, the social nature of reality is emphasized and it is suggested that a significant loss might be viewed as a *loss of self* (Charmaz 1980). Moreover, the symbolic interactionist view on emotions strengthens the concept of grief as an *emotional role* and the importance of feeling rules and a normative regulation of grief (for a more detailed discussion about grief and symbolic interactionism see Jakoby 2012).

First, we have to think about the meaning of loss (Charmaz 1980; Lofland 1985). It is not only the loss itself; death shakes the foundations on which the self of the survivor is constructed and known. The death of a significant other means a loss of self-image and subjective world and a symbolic death of ourselves (see Charmaz 1980: 297). From the viewpoint of Mead (1965), the self is constructed and defined in social interaction with significant others. Representations of the loved one are part of the person's self-concept (Archer 1999). When the "other" dies, the social nature of the self becomes painfully obvious and makes grief a social phenomenon. It is not only the loss of the loved one but also the loss of self that was constructed through interactions with the deceased (see Bradbury 1999: 175; Valentine 2008: 97). From this perspective, grief is defined as a painful reconstruction or rebuilding of the self and everyday life (Schmied 1985). The sociological concept of *threads of connectedness* (Lofland 1985) best describes the multi-dimensional connections that are destroyed by death. Understanding grief sociologically means that we have to understand the social bond between the deceased and the survivor and the variation in significance that others may have for us. We are linked to others by the roles we play, the support we receive, the wider networks others make available to us, the selves others create and sustain, the reality they validate or the future they make possible for us (see Lofland 1985: 175).

Within symbolic interactionism, grief is also conceptualized as an *emotional role*. According to Hochschild (1979, 1983), emotions are guided by feeling rules. Feeling rules are scripts for emotions in a given society and culture. They are social norms and specify the emotions that individuals should feel or express in a given situation. Feeling rules govern the intensity and duration of grief, whereas display rules refer to the expression of grief. In sum, Hochschild (1983) emphasizes the importance of the grief role: "The ways in which people think they have grieved poorly suggest what a remarkable achievement it is to grieve well" (Hochschild 1983: 68). Several "misfits" are possible: We can grieve too much or too little. In this case, we "overmanage or undermanage grief" (see Hochschild 1983: 64). There can be a "misfit in timing" when grief emerges too late or a "misfit in placing". Only the funeral and the cemetery are considered to be the right places to express grief and sadness, whereas other places, such as the workplace, are inappropriate (see Hochschild 1983). Implicit feeling rules can be found in psychological classifications of pathological grief. The labels of "inhibited grief", "delayed grief", or "chronic grief" (see Stroebe and Stroebe 1987) represent deviances in contrast to the model of normal grief. They contain feeling rules that inform about the expected duration and intensity of grief (see also Charmaz and Milligan 2006: 530). For example, books written for bereaved individuals or counseling literature include these guidelines for normal grief (Wortman and Silver 2007).

The sociological conceptions of disenfranchised grief (Doka 2002) and demoralized loss (Fowlkes 1990) are linked with sociological theories of emotions because both take into account the importance of feeling rules. The right to the expression and feeling of grief varies according to the social comprehension of a legitimate relationship and loss (see Fowlkes 1990: 649). Fowlkes (1990) emphasizes the social regulation of grief as it follows from a differential valuation of relationships. This standard is visualized by the topics of most of the bereavement research which deal primarily with the death of a nuclear family member and neglect relationships such as colleagueship as well as stigmatized losses (e.g., partner in a homosexual relationship, extramarital affair; see Fowlkes 1990; Doka 2002). These losses are unrecognized, not acknowledged or publicly mourned because the survivors lack the right to grieve (see Charmaz and Milligan 2006: 529). According to Hochschild (1979, 1983), the deviant expression of grief will be managed or even suppressed. When feelings deviate from feeling rules, individuals can do emotion work behaviorally, bodily, or cognitively (see Hochschild 1979: 562). It is important to pay attention to family norms about the right way of grieving and pressure by family members about how to feel and behave (see Walter 2005). Besides cultural norms of grieving, these "lay and semi-psychiatric notions of abnormality" (Walter 2005: 75) affect grief. Goodrum (2008) refers to emotional socialization in everyday life and reveals the complexity of strategies and techniques, such as avoiding the topic, saying "It is time to move on", or crying as a reflection of the inability to handle the news or offer assistance (see Goodrum 2008: 430). The most common strategies of grief management aimed at bringing emotions into line with normative expectations are restraining grief and pretending to feel good. Whether it is the management of one's own emotions in terms of deep acting or playing down grief as a display act (Hochschild 1983) to reduce others' discomfort, for instance, suffering relatives, and to prevent them from also being sad (see Goodrum 2008: 435).

CHARACTERISTICS OF THE ONLINE WORLD

The world of the Internet differs from the real world in many respects. The most obvious differences result from the fact that the online world completely lacks any physical qualities (Geser 1998a, 2002). This particularly refers to time, space, and matter. Given the absence of these defining features of reality, the Internet has "less stabilizing and complexity-reducing effects" (Geser 1998a: 2). In the online world, there are no spatial distances; any place can be reached expending the same effort in terms of time and cost. Time is a factor that matters in yet another dimension. Every user can decide individually when to post on a forum or whose messages he or she chooses to read. In the real world, only one person can speak at a time while all others are expected to listen. The Internet clearly offers the advantages of not relying on simultaneous physical presence, providing the opportunity for synchronous communication, and facilitating reference to previous messages by making data permanently available (Geser 1998a, 2002). The online world provides users with the opportunity to adopt different roles that to some degree may compensate for and/or conflict with aspects of their lives in the real world. In addition, the online world lacks the perceptibility of objects and subjects. As online communication is not aided and structured by visible gestures and facial expressions, we must construe our counterparts' emotions and opinions from their written messages only (Geser 1998a, 2002). In the web, the act of expressing emotions takes more time, so that emotions are uttered in a more controlled and less spontaneous manner than in the real world (Derks et al. 2008). The communication in the online world is no less emotional than in the real world. Quite the contrary, especially negative emotions are expressed even more overtly in computer-based communication. This phenomenon can mainly be attributed to user anonymity resulting from not being socially present and thus not having to fear any negative sanctions (Derks et al. 2008). Moreover, one can observe a tendency toward emotionalizing communication while following strictly formalized rules. With respect to community building, membership in a social network is emphasized by the extensive use of emotionalized language (Geser 1998a). In the real world, belonging to a particular social group is already expressed by spending time together, but in virtual communication, this sense of affiliation has to be created by formulating, and that means writing, clear and explicit emotional statements. Moreover and contrary to expectations, the Internet does not support the development of a more subjective and individualized language; users rather tend to stick to the language norms of the real world (Geser 1998a).

The lack of physical qualities "increases the need for establishing endogenous structures" (Geser 1998a: 1). Factors that do not depend on physical qualities thus gain major significance, such as values, reputations, and other common ground shared by Internet users. Internet users also resort to already existing structures, be they concepts or symbols. Owing to the fact that they cannot see or hear each other, clearly comprehensible words and signs are key in rendering communication unequivocal and intelligible to all. On the Internet, weblinks, images, videos, and texts can be incorporated into a single digital file. This enables websites to simultaneously interact with multiple communication channels (see Hein 2009: 37). If we consider Geser's (1998a) line of argument, we may hold that the reason for employing hypermedia strategies on the Internet is to make up for the lack of structure by conveying emotions via images and videos. Due to the wide range of options for selecting and modifying content, the Internet displays high levels of interaction and represents a platform for combining the most diverse media into a single one. These characteristic features of the Internet and the absence of physical qualities are increasingly influencing funeral culture.

VIRTUAL CEMETERIES AS A FIELD FOR RESEARCH

The first virtual cemeteries can be traced back to the year 1990 (Spieker and Schwibbe 2005). Over the last two decades, the range of online cemeteries has increased in number and diversity. The benefits of web memorials are flexible timing, easy access, permanent visibility and sharing with others without the restrictions of space and time (Roberts and Vidal 1999–2000; Geser 1998b, 1998c). Besides the savings in cost and time and lifelong availability, modifiability is another factor that distinguishes the online cemetery from its traditional counterpart. The virtual grave can be modified according to the emotional state of the bereaved and visitors and is extremely malleable compared to the static memorial (see Spieker and Schwibbe 2005: 235).

One of the Internet's key functions is to provide people access to information and communication worldwide without much effort in terms of time and cost (Geser 1998c), especially in light of the fragmentation of modern lives (Walter 1996). Memorial sites are particularly useful for the remembrance of mobile individuals who have left highly dispersed and incoherent social networks of friends and family. In addition, web memorials offer people whose grief is not accepted in mainstream society a way to express their feelings of loss, whether the deceased was a gay lover, an ex-husband, or a pet (De Vries and Rutherford 2004). Therefore, web memorials are best described as a platform for recognizing *disenfranchised grief* (Doka 2002). Moreover, the model of *continuing bonds* (Walter 1996), which conceptualizes grief as a process of talking about the dead, is represented in the web memorials (De Vries and Rutherford 2004). However, the existing studies do not explicitly refer to the theoretical assumptions of symbolic interactionism. Our analysis will extend previous research on web memorials by focusing on the concept of feeling rules.

RESULTS[1]

This study concentrates on the following research questions: (1) Can the feeling rules that determine the expression of grief in the offline world be identified in the online world as well? (2) Does the Internet give rise to a new form of Grief 2.0? In particular, we want to investigate how the absence of physical parameters in the online world, such as space, time, and matter, affect the way grief is expressed. Web memorials represent a "thanatological resource" (Sofka 1997) for studying the emotion of grief, similar to condolence books (Brennan 2008). Web memorials provide insight into the social expression of grief and therefore offer a fruitful research tool for the sociology of emotions in general and bereavement research in particular. Our analysis is based on qualitative content analysis (Mayring 1997) using inductive and deductive categories. The sample consists of two German websites:

> *Memorta* (www.memorta.com), and
> *Strasse der Besten* (www.strassederbesten.de)

In 2011 both web memorials were free of charge and included demographic information of the deceased (name, and date of birth and death). This analysis is based on publicly available memorials. We focus on the posted messages to the deceased, and narratives about the deceased as well as guestbook entries.

Although the grieving process is highly individualized in modern societies, feeling rules govern the expression of grief as postulated by symbolic interactionism (see earlier discussion). One user even explicitly refers to the feeling rules that apply in the offline world:

> Maybe it is better that way; they ask me, [. . .] "Are you feeling better?" They ask and just want me to be the person I was before the death of my child.
> (http://www.strassederbesten.de/onlinefriedhof/virtueller_friedhof_grab_2389.html)

And:

> The advantage of an online cemetery to me is that I can mourn alone and in peace. Where I don't feel watched but feel understood. [...] Among the people I know [...] , there are some who think that after a certain period of mourning it is enough. Who don't understand that you always mourn. You will mourn a lifetime.[2]

We identify four themes centered on the *freedom from feeling rules* in virtual cemeteries: (1) the absence of any limitations of time, (2) spiritualization and afterlife beliefs, (3) a new model of "the family," and (4) talking to the dead.

Theme 1: Absence of Time Limitations

The freedom from time restrictions and the expression of ongoing grief without seeking recovery are typical of the virtual expression of grief. We only found one message in the guestbook at *Memorta* that directly addressed the feeling rule of recovery. An anonymous guest pointed out that "time will also heal your wounds".[3] For the most part, however, mourning can go on for a wide range of different time periods after a death occurs, which contradicts the classic stage model of grief or the concept of "normal grief".

> Still in our heart, even after three years.
> (http://www.memorta.com/internetfriedhof/memo.php?id=339)

> It has been 30 years now since you passed away. But it still feels like it happened yesterday.
> (http://www.strassederbesten.de/onlinefriedhof/virtueller_friedhof_grab_3691.html)

> My heart will always be in grief, whether on good days or bad days.
> (http://www.strassederbesten.de/onlinefriedhof/virtueller_friedhof_grab_2389.html)

And:

> Dear Mom, it's been 17 years now that you are not with us anymore. I miss you so.
> (http://www.strassederbesten.de/onlinefriedhof/virtueller_friedhof_grab_13646.html)

Therefore, a virtual cemetery represents a "cultural innovation because it has the potential of providing a focus for long-term mourning" (Geser

1998c: 10). The social and emotional relationship with the deceased does not end even after years. Web memorials provide a means for expressing long-term emotional processes that have long existed at the psychological level but had no means of finding social expression in the offline world due to a lack of conventional institutions or religious rituals or because of the restrictions governing face-to-face interactions (see Geser 1998c: 14).

Theme 2: Spiritualization and Afterlife Beliefs

For the survivors, death is not the end of the emotional and social relationship with the deceased. Web memorials reveal the social construction of death as *immortality* and spiritual and afterlife beliefs best described as heaven, paradise, angels, reunion, peace, and release. Web memorials include messages addressed to the deceased, such as "see you in heaven"[4] or "say hello to the angels".[5] The "individualized spirituality" is based on religious discourse and religious terminology—a finding supported by Brennan (2008).

> When you lose a loved person, you gain a guardian angel.
> (http://www.memorta.com/internetfriedhof/memo.php?id=168)

And:

> I am sure that you are an angel now. [. . .]
> My love, I hope you found your paradise.
> Enjoy the calmness, the light, and the peace. [. . .]
> I will be so happy when we see each other again.
> (http://www.memorta.com/internetfriedhof/memo.php?id=95)

Theme 3: A New Model of "The Family"

Emotional bonds are diverse and complex: "It is not the category of relationship loss that occasions grief but, rather, the loss of significant attachment" (Fowlkes 1990: 638). A look at web memorials reveals that the feelings of loss are not only restricted to the members of the traditional nuclear family. Instead, web memorials display a new conception of family and a new diversity of legitimate losses. They also provide the extended family with an opportunity to express grief. In web memorials, one finds messages from and memorials for various categories of kin, not only members of the nuclear family or along the lines of generational relations with grandparents. The new model of "the family" is represented by virtual family tombstones and kinship memorials.[6] Web memorials also refer to family reunion after death and family gatherings on Christmas, birthdays, or anniversaries. Moreover, friends have the same status as family members in expressing grief and writing about their memories and feelings. These alternative constructions of family in web memorials emphasize the idea of friends as family, the importance of the extended family and reconfigured kin networks, or some combination of biological and social kin. Especially relationships with so-called "distant kin"—such as aunts and uncles, or cousins—are of great importance for the social construction of the family.

Theme 4: Talking to the Dead

The person is dead, but the social and emotional relationship still exists. Web memorials support the continuation of a dialogue with the deceased by providing a forum for talking to them (Brennan 2008). Whereas talking *about* the dead is socially accepted in offline situations, talking to the dead is traditionally associated with pathological grief reactions. In the medical model, manifestations of continuing grief, such as sensing the presence of the dead or talking to the dead, would be labeled as pathological (Wortman and Silver 2007; Klass and Walter 2007). Nevertheless, studies show that the cemetery is a common place for conversation with the dead (Silverman and Nickman 1996; Francis et al. 1997). Moreover, talking to the dead can be part of counseling practice or grief therapy (for a review, see Klass and Walter 2007).

> Hi, my Big Boy. We just came home from our vacation and I am just visiting you here. I will come to the cemetery tomorrow.
> Love, Mom.
> (http://www.memorta.com/internetfriedhof/memo.php?id=170)

> Hello Mom. Today is 02/23/2011, and I visited Grandma Grete in the hospital yesterday; things do not look good at all. According to the doctors, it will only another 3 to 4 days and she will be with you.
> (http://www.memorta.com/internetfriedhof/memo.php?id=85)

Survivors address their loved ones on special occasions, such as anniversaries, birthdays, Christmas, or family gatherings; for example, they may wish them "Happy Birthday!" just like in the real world.[7] Furthermore, the ongoing conversation with the dead via virtual cemeteries may also aim at resolving open issues and in this sense "finishing" the relationship; this may involve the desire to talk about misunderstandings and quarrels or seek reconciliation:

> Many things did not turn out as we or you hoped. No longer shall it bother us.
> (http://www.memorta.com/internetfriedhof/memo.php?id=96)

The Internet, through its medium of virtual cemeteries, enables establishing a continuous virtual conversation with the deceased. Apart from traditional cemeteries, web memorials enhance the social bonds between the living and the dead. Talking to the dead is now separated from the dead body and its territorial location in the grave at the cemetery (Geser 1998b, 1998c).

Virtual cemeteries provide *virtual support* for the feelings of the bereaved. The posted messages of friends, family, kin, or even unknown persons in guest books serve as a "system of understanding", whether in the form of a direct response to the bereaved or a personalized response to the loss by displaying some form of empathetic understanding or pity as the following quotations reveal.

> You are allowed to permit pain and grieving.
> (http://www.strassederbesten.de/onlinefriedhof/virtueller_friedhof_grab_2389.html)

And:

> They don't tell me, "You've mourned enough now." They say, "Of course you have the right to mourn, of course you will mourn all your life."
> (http://www.3sat.de/page/?source=/kulturzeit/themen/152107/index.html)

The survivors honor the emotional support via Internet communication, for instance, by thanking the stranger who lit a candle on the remembrance site or by expressing their appreciation for the general support received. For example:

> With these lines we warmly thank all those who lent their support during a difficult time by expressing their condolences or by entries in the guestbook. Your support has been very comforting and will continue to be a source of solace to us in the future.
> (http://www.memorta.com/internetfriedhof/memo.php?id=95)

Consequently, web memorials are described as a virtual "self-help movement" (see Spieker and Schwibbe 2005: 239). Virtual cemeteries can also be characterized as *systems of textualized emotions*, similar to self-help groups, which Walter (1994) calls "systems of listening".

CONCLUSIONS

Virtual cemeteries serve as a "platform for the social expression of grief" (Brennan 2008: 326), a continuous forum for the expression of grief in general and writing down one's emotions in particular. Moreover, web memorials help bridge the gap between language and loss, the living and the dead, between private and public mourning (Brennan 2008). Therefore, the Internet enables a new post-modern discourse about grief and death (see Geser 1998c: 25). In this study, the Internet served as a "methodological laboratory" (Geser 1998a) to investigate the sociological concept of feeling rules as postulated by symbolic interactionism (Hochschild 1983). Compared to the circumstances characterizing the offline world, which are mostly described in terms of "policing grief" (Walter 1999), different feeling rules apply in virtual cemeteries with regard to time limitations, spiritualization and afterlife beliefs, who counts as "family", and talking to the dead. The expressions are unaffected by pressures of "social desirability" (Geser 2002: 25). There is no social pressure to perform "normal grief".

The Internet compensates functional deficits of the "offline society" by providing opportunities for the expression of long-term grief and spirituality, including a wider range of social relationships in the expression of legitimate grief, and offering a forum for talking to the dead, which usually occurs hidden in the privacy of individual minds or interpersonal relationships (see Geser 1998c: 10, 19). These "deficits" are rooted in a lack of institutionalized answers for dealing with long-term grief, in religious and cultural rituals that insufficiently address such needs, or the inability of social relationships to deal with the suffering and public expressions of grief on part of the survivors. Studying online grief is useful for gaining a new understanding of conventional offline grief and its constraints for the bereaved. The Internet expands the range of expressions by providing means for explicating thoughts never spelled out under conventional conditions of daily life (see Geser 1998c: 10, 19).

In the online world, the emergence of so-called "communicative islands" provides orientation by focusing on specific topics and functions (see Geser 1998a: 7). The virtual cemetery can be characterized as such a "communicative island". Of course, it only represents an exclusive and small fraction of the entire online world with its specialized function of expressing grief and love. It would therefore be inadequate to simply equate the online cemetery with the entire online world. Hence, our conclusion that no feeling rules exist is valid for online cemeteries but *not* for the online world in its entirety. Moreover, a similarity to physical cemeteries must be emphasized. Expressions of joy, happiness, or contentment unrelated to grief are hardly found in web memorials. As explained earlier, there can be a "misfit in placing" the emotion of grief (Hochschild 1983). This aspect becomes more evident when one compares the online with the real cemetery. Cemetery behavior—a gap in bereavement research on the practices in Western societies—is characterized by social norms prohibiting certain behavior in this setting, such as running, speaking loudly, or laughing. Such norms constitute, in Hochschild's words (1979), "display rules". The same can be observed on the Internet where the online cemetery is defined as a place of mourning by the feeling rules for expressing grief. The statements "use this platform to mourn" (www.memorta.com) and "don't let our loved ones be forgotten" (www.strassederbesten.de) can be interpreted as feeling rules directed at the bereaved and requesting them to mourn and express their emotions on these particular websites. This finding is supported by the criticism of the continuing bonds model and the general concern that this model may become the new paradigm (Howarth 2007). Moreover, the model does not take into account that some people want to forget the dead: the dead are not always welcome (see Howarth 2007: 213).

In light of these facts, one can hypothesize that the feeling rules prevail in the online world as well as in the real word. This confirms Geser's (1998b, 1998c) hypothesis that in the online world one can observe sociological processes and assess the validity of theories in an ideal-type manner. For future research projects, it would be interesting to investigate whether the freedom of expressing grief is restricted to virtual cemeteries exclusively or whether the social expression of grief is also encountered in virtual social networks, such as Facebook.

Has the Internet given rise to a new form of Grief 2.0? The answer to this question is no. The feeling rules exist prior to the medium by which emotions are expressed. It is not the medium that changes the individual feelings of grief. The Internet only influences the way emotions are expressed but not the experience of the emotion itself. For example, in the online world negative emotions are expressed more often than in the real world (Derks et al. 2008). It would be a circular argument to conclude from this that negative emotions are more frequent in the online world. One explanatory variable to account for this observation is the anonymity in the Internet, which allows expressing negative emotions without fear of sanctions by others (Derks et al. 2008). Although in web memorials the expression of emotions is subject to the laws of the online world, the emotions expressed are not altered by the medium. It is not the medium but the cemetery, as a place defined by the feeling rule for expressing grief and mourning, that determines the private emotion of grief. Although private and intimate emotions are revealed, web memorials refer to "highly standardized topics and conventional codes of expressions" (Geser 2002: 20) and a "pre-existing conservative structure of meaning" (Brennan 2008: 336). These standardized codes of grief are represented by tombs, candles, flowers, photos, eulogia, or secular poems which represent symbols of the "objective culture" of mourning (Geser 2002: 20).

The emergence of virtual cemeteries represents a major change in grief culture. It is not a singular phenomenon but an ongoing process. The European history of the cemetery has always been subject to changes of various sorts: the decline in family graves and the emergence of the Christian cemetery, the relocation of cemeteries outside the city walls and away from the church, or the introduction of cremation and hence of the urn grave (Ariès 2009). To mention a more current example, the governments in Germany and Switzerland both are in the process of loosening the rules for the individualization of grave stones. All these examples show that society always creates spaces where grief has its place in accordance with the respective social, cultural, and historical circumstances of the time. What matters is not where this place is located but what feeling rules prevail.

NOTES

1. The statements cited from the web were translated from German into English (retrieved June 1, 2011).
2. http://www.3sat.de/page/?source=/kulturzeit/themen/152107/index.html, retrieved June 1, 2011.
3. http://www.memorta.com/internetfriedhof/memo.php?id=183
4. http://www.memorta.com/internetfriedhof/memo.php?id=192
5. http://www.memorta.com/internetfriedhof/memo.php?id=31
6. http://www.strassederbesten.de/onlinefriedhof/virtueller_friedhof_grab_7165.html
7. http://www.memorta.com/internetfriedhof/memo.php?id=192

REFERENCES

Archer, J. (1999) *The Nature of Grief: The Evolution and Psychology of Reactions to Loss*. London and New York: Routledge.

Ariès, P. (2009) Geschichte des Todes (12th ed.). München: Deutscher Taschenbuch Verlag.

Bradbury, M. (1999) *Representations of Death. A Social Psychological Perspective*. London and New York: Routledge.

Brennan, M. (2008) Condolence Books: Language and Meaning in the Mourning for Hillsborough and Diana. *Death Studies*, 32, 326–351.

Charmaz, K. (1980) *The Social Reality of Death*. Reading: Addison-Wesley.

Charmaz, K. and Milligan, M.J. (2006) Grief. In J.E. Stets and J.H. Turner (Eds.), *Handbook of the Sociology of Emotions* (pp. 516–543). New York: Springer.

Cochran, L. and Claspell, E. (1987) *The Meaning of Grief. A Dramaturgical Approach to Understanding Emotion*. Westport, CT: Greenwood Press.

Derks, D., Fischer, A.H. and Bos, A.E.R. (2008) The Role of Emotion in Computer-Mediated Communication: A Review. *Computers in Human Behavior*, 24, 766–785.

De Vries, B. and Rutherford, J. (2004) Memorializing Loved Ones on the World Wide Web. *Omega*, 49, 5–26.

Doka, K.J. (Ed.). (2002) *Disenfranchised Grief*. Champaign, IL: Research Press.

Fowlkes, M.R. (1990) The Social Regulation of Grief. *Sociological Forum*, 5, 635–652.

Francis, D., Kellaher, L. and Lee, C. (1997) Talking to People in Cemeteries. *Journal of the Institute of Burial and Cremation Administration*, 65, 14–15.

Geser, H. (1998a) *Metasoziologische Implikationen des „Cyberspace". Sociology in Switzerland: Toward Cybersociety and Vireal Social Relations*. Zürich. Retrieved January 6, 2011 from http://socio.ch/intcom/t_hgeser03.htm

Geser, H. (1998b) Yours Virtually Forever. Elektronische Grabstätten im Internet. In K. Imhof and P. Schulz (Eds.), *Die Veröffentlichung des Privaten. Die Privatisierung des Öffentlichen* (pp. 120–135). Opladen: Westdeutscher Verlag.

Geser, H. (1998c) *Yours Virtually Forever. Sociology in Switzerland: Toward Cybersociety and Vireal Social Relations*. Zürich. Retrieved January 6, 2011 from http://socio.ch/intcom/t_hgeser07.htm

Geser, H. (2002) *Towards a (Meta-)Sociology of the Digital Sphere. Sociology in Switzerland: Toward Cyberspace and Vireal Social Relations*. Zürich. Retrieved January 6, 2011 from http://socio.ch/intcom/t_hgeser13.pdf

Goodrum, S. (2008) When the Management of Grief Becomes Everyday Life: The Aftermath of Murder. *Symbolic Interaction*, 31, 422–442.

Hein, D. (2009) *Erinnerungskulturen online. Angebote, Kommunikatoren und Nutzer von Websites zu Nationalsozialismus und Holocaust*. Konstanz: UVK Verlagsgesellschaft.
Hochschild, A.R. (1979) Emotion Work, Feeling Rules and Social Structure. *American Journal of Sociology*, 85, 551–575.
Hochschild, A.R. (1983) *The Managed Heart. Commercialization of Human Feeling*. Berkeley, CA: University of California Press.
Howarth, G. (2007) *Death and Dying. A Sociological Introduction*. Cambridge: Polity Press.
Horwitz, A.V. and Wakefield, J.C. (2007) *The Loss of Sadness. How Psychiatry Transformed Normal Sorrow Into Depressive Disorder*. Oxford: Oxford University Press.
Jakoby, N. (2012) Grief as a Social Emotion. Theoretical Perspectives. *Death Studies*, 36, 679–711.
Klass, D. and Walter, T. (2007) Processes of Grief: How Bonds are Continued. In M.S. Stroebe, R.O. Hansson, W. Stroebe, and H. Schut (Eds.), *Handbook of Bereavement Research. Consequences, Coping, and Care* (pp. 431–448). Washington, DC: American Psychological Association.
Lofland, L.H. (1985) The Social Shaping of Emotion: The Case of Grief. *Symbolic Interaction*, 8, 171–190.
Mayring, P. (1997) *Qualitative Inhaltsanalyse. Grundlagen und Techniken* (6th ed.). Weinheim: Deutscher Studien Verlag.
Mead, G. H. (1965) *Mind, Self & Society* (13th ed.). Chicago and London: The University of Chicago Press.
Parkes, C.M. (1972) *Bereavement. Studies of Grief in Adult Life*. London: Tavistock.
Roberts, P. and Vidal, L.A. (1999–2000) Perpetual Care in Cyberspace: A Portrait of Memorials on the Web. *Omega*, 40, 521–545.
Schmied, G. (1985) *Sterben und Trauern in der modernen Gesellschaft*. Opladen: Leske+Budrich.
Silverman, P.R. and Nickman, S.L. (1996) Children's Construction of Their Dead Parents. In D. Klass, P.R. Silverman, and S. Nickman (Eds.), *Continuing Bonds: New Understandings of Grief* (pp. 73–86). Washington, DC: Taylor & Francis.
Sofka, C.J. (1997) Social Support 'Internetworks,' Caskets for Sale, and More. Thanatology and the Information Superhighway. *Death Studies*, 21, 553–574.
Spieker, I. and Schwibbe, G. (2005) Nur Vergessene sind wirklich tot. Zur kulturellen Bedeutung virtueller Friedhöfe. In N. Fischer and M. Herzog (Eds.), *Nekropolis: Der Friedhof als Ort der Toten und der Lebenden* (pp. 229–242). Stuttgart: Kohlhammer.
Stroebe, W. and Stroebe, M.S. (1987) *Bereavement and Health: The Psychological and Physical Consequences of Partner Loss*. Cambridge: Cambridge University Press.
Valentine, C. (2008) *Bereavement Narratives: Continuing Bonds in the 21st Century*. London and New York: Routledge.
Walter, T. (1994) *The Revival of Death*. London/New York: Routledge.
Walter, T. (1996) A New Model of Grief: Bereavement and Biography. *Mortality*, 1, 7–25.
Walter, T. (1999) *On Bereavement. The Culture of Grief*. Maidenhead: Open University Press.
Walter, T. (2000) Grief Narratives: The Role of Medicine in the Policing of Grief. *Anthropology & Medicine*, 7, 97–114.
Walter, T. (2005) What is Complicated Grief? A Social Constructionist Perspective. *Omega*, 52, 71–79.

Winkel, H. (2001) A Postmodern Culture Grief? On Individualization of Mourning in Germany. *Mortality*, 6, 65–79.
Wortman, C.B. and Silver, R.C. (2007) The Myths of Coping With Loss Revisited. In M.S. Stroebe, R.O. Hansson, W. Stroebe, and H. Schut (Eds.), *Handbook of Bereavement Research. Consequences, Coping, and Care* (pp. 405–429). Washington, DC: American Psychological Association.

5 Islamic Emoticons
Pious Sociability and Community Building in Online Muslim Communities

Andrea L. Stanton

Since the late-1990s, the communications field has seen a growing body of research on computer-mediated human interaction that has focused on the role and importance of emoticons.[1] Since the early-2000s, the fields of Middle Eastern and Islamic studies have seen a growing corpus of research on the role and importance of the Internet for Muslim life in the 21st century—whether for personal communications, work, piety, or learning. Yet there has been little crossover between these fields—particularly surprising given Middle East and Islamic studies' research focus on web boards and chat forums. The major scholars in the field, Gary Bunt and Jon Anderson, have said almost nothing about the smiley faces, modified smileys, and text-based icons that pepper forum discussions.[2] Even scholars who include emoticons in the transcripts of their research, as Roxanne Marcotte has done in her studies of online Australian Muslim communities, offer few comments either about their meaning or the work they do within the forum discussion.[3]

This chapter seeks to draw from the research done in both areas, using it as a foundation to probe the way emoticons are produced, used, discussed, and understood among Muslims online. It maps the field of Islamic emoticons, outlining the ways in which this field is being defined, theorized, and debated within online Muslim communities. It begins with a literature review, considering both communications work on emoticons in general and Middle Eastern and Islamic studies research on Muslim communities and their use of the Internet for pious purposes. It provides a brief history of emoticons, grounding current applications in emoticons' original use as a communication work-around—a compensation for the emotional ambiguity of textual communications.

Turning to Islamic emoticons, this study focuses first on the endorsement of particular emoticons by Islamic websites, and then turns to broader Muslim debates about emoticons in general. These debates range from questioning the acceptability of any visual media, to questions about the inadmissibility of 'secular' emoticons on Islamic chat sites. These debates help put in question the presumed relationship between modernity and secular liberalism by asking how the Internet fits into this relationship. Rather than adopting the classic Habermasian model of the public sphere as secular and emotionless, it recognizes the role of faith and affect in the digital public sphere by focusing on the emoticons used on Islamic website forums. It argues that these emoticons reflect the emergence of new emotional practices by users, who generally self-identify as practicing Muslims. They help ameliorate the anonymity and disembodiment of online avatars, contributing to community building, while also reflecting a deeper cultivation of self. At the same time, the active role that users play in pushing for site administrators to enable Islamic emoticons and disable smileys and other generic emoticons reflects the empowerment of the ordinary Muslim and broadening definitions of religious authority characteristic of the modern era.

This study concludes by emphasizing the need to examine the work that these emoticons do: unlike the generic smileys of yesteryear, today's emoticons, it argues, reflect the personal expression of particular communities—in this case, practicing Muslims. Yet they do more than reflect: as scholars like Saba Mahmood have noted in similar contexts, they encourage forum members in the ongoing process of pious self-cultivation, providing a visual reminder of the particular sensibilities endorsed within these online community spheres. As a result, they shift what Mahmood has termed the "piety movement"[4] from the world of face-to-face interactions and embodied experience into the virtual sphere—a virtuality intensified by the dispersed locations of site users.

LITERATURE REVIEW

Emoticon research began in the mid-1990s, when researchers began grappling with the terminology and function of the smiley faces, frowns, and other text-based images that Internet users were creating to aid in computer-mediated communicating. Early research focused on email communications, noting that this was a user-driven phenomenon rather than one created top-down by computer programmers.[5] As instant messaging and other conversation formats emerged, researchers began including these formats in their studies. Most emoticon research to date has fallen into the fields of either communications or computer science. In both fields, most studies have worked to assess both how emoticons work—the role they play in online communication—and also to determine how well they work—that is, the impact they have on that communication.

Much communications and computer science research begins from the premise that in a text-based communication environment such as the Internet, emoticons play an important role in filling in a gap that in direct interaction—whether in person or over the phone—would be filled by visual, aural, and/or physical cues. As Eli Dresner and Susan Herring explain in a 2010 article on non-verbal aspects of computer-mediated communication (CMC), "The term 'emoticons'—short for 'emotion icons'—refers to graphic signs, such as the smiley face, that often accompany computer-mediated communication. They are most often characterized as iconic indicators of emotion, conveyed through a communication channel . . . parallel to the linguistic one."[6] Dresner and Herring argue that, in fact, emoticons do more than indicate emotion: they also indicate that the words they accompany are intended to have an illocutionary force, or to function as a kind of speech act. While not all scholars define emoticons this way, it might be helpful to think of emoticons as both standing in for nonverbal communication that computer mediation blocks and as highlighting or reinforcing the intent of particular computer-mediated statements.

A number of recent studies have focused on assessing the impact of emoticons on conversation, asking whether they compensate adequately for nonverbal communication in non-mediated contexts, and whether they affect mediated communication in other ways. These studies have a highly international component, with research as likely to be conducted in Asia or Europe as in North America. A 2008 Dutch study involving high school students argued that emoticons do effectively fill the role of nonverbal communication, noting that emoticons "strengthen the intensity of a verbal message" when used to reinforce that message, but also "create ambiguity and express sarcasm" when used to undercut the verbal message.[7] In another article published by the same team of scholars, they noted that individuals engaging in online conversations primarily used emoticons to communicate emotion, to strengthen verbal messages, and for humor. Emoticon usage correlated directly with how well the individual knew his or her conversation partner: individuals used more emoticons when communicating with friends than with people they knew less well. Further, emoticon usage skewed positive: individuals were less likely to use emoticons to convey a negative message.[8]

Children were also the subjects in a recent Taiwanese study by Fang-Wu Tung and Yi-Shin Deng on the use of static and dynamic emoticons as "social cues" for children in computer-aided education, which concluded that the use of emoticons helps children engage with the computer and increases their motivation to learn.[9] Some recent scholarship has argued that since emoticons have developed into a standard mode of computer-mediated communication, they provide both verbal and non-verbal cues—leading one scholar to propose renaming them "quasi-nonverbal" cues.[10] On the other hand, one recent psychology study has found that users demonstrate a widely varying ability to correctly interpret emoticons, suggesting that even the most frequently used emoticons may have a range of interpretive meanings and uses.[11]

Little computer science-related research appears to have considered Muslims as the particular object of study; on the other hand, there appears to have been similarly minimal research focused on other particular religious groups. There has, however, been considerable research on Islam and the Internet, and particularly on the various ways Muslims interact online.

One of the pioneers in the field has been Jon Anderson, an anthropologist whose work on Muslims and the Internet dates to the late-1990s, when he produced a groundbreaking edited book, *New Media in the Muslim World*, with Dale Eickelman.[12] Early scholarship such as this collection often focused on communication and network building among Muslims online, particularly in diasporan or expatriate contexts. This work suggested that new media might in turn foster the creation of new publics with a more active relationship to their media consumption—laying the groundwork for a new public sphere. Further, early works like Anderson and Eickelman's, suggested this new public and its new public sphere would put pressure on the largely authoritarian states of the Muslim-majority world.

One strain of scholarship has taken up the question of the relationship between new media, new publics, envisioning the Internet fostering new relationships between citizens and state power. Some scholars have applied various social science theories, as in the case of a recent study examining whether social movement theory could help explain the role played by the Internet in the 2005 Kifaya Movement in Egypt and the "Green Revolution" in Iran. This study concluded that the web supports rather than hinders social groups and their mobilization for "real world" action.[13] The question of new publics and their ability to foster broader change has remained an open one as recently as 2010, when an Australian study argued that second-generation Internet features (in particular social media) have played a crucial role in the evolutionary development of Muslim publics, leading over time to a broad transition away from authoritarianism.[14] On the other hand, some scholars have suggested a possible connection between Internet activity and political Islam, or about mobilizing Muslims for faith-based activism in the political realm.[15]

A second strain of scholarship since the mid-2000s has focused on the cultivation and maintenance of piety through educational websites, clerics' websites and fatwa engines, and Islamic forums. While many studies have focused on Muslim communities exclusively, some have targeted Muslim communities in the context of other religious groups, as with a recent study on the Internet and religion in South and Southeast Asia.[16] One key finding of various studies has been a clear indication that Muslims online are motivated by a personal sense of piety—or a desire to know more about their religion—and find websites and forums more compelling than the traditional clerical loci of Islamic authority (like shaykhs, imams, and `ulama).[17] Yet the forum threads reprinted in several studies indicate on the other hand a clear longing for authority, as forum posters complain that the forum moderators are too passive, or lament that posters are all ordinary Muslims, with no one able to provide a clerical perspective.[18] This longing has received little attention from scholars.

As new forms of online engagement have emerged, research has shifted to accommodate them. For example, the emergence of podcasts as a locus for religious scholars and imams, as well as others interested in either claiming authority to interpret Islam or sharing their knowledge, has been tracked in several recent articles, most notably Jan Scholz, Tobias Selge, Max Stille, and Johannes Zimmermann's "Some Remarks on the Construction of Religious Authority in Islamic Podcasts."[19] Few scholars have paid much attention to the telecommunication laws and other government strictures helping to both shape and also constrain the way that Muslims in various countries use the Internet. However, a recent study on the "geographies" of Muslim-majority countries has done much to fill the gap, arguing both for the important role of policies and infrastructure and for the increasing significance of e-commerce and "e-government."[20] The major scholar of piety and pious interactions online, however, has been Gary Bunt, whose work focuses on what he terms the "Cyber-Islamic environment."[21]

A BRIEF HISTORY OF THE EMOTICON

Emoticons developed as a 'work-around' for the exclusively textual nature of the early Internet. Scholars offer differing dates for the "first" emoticon, with some citing 1979 and others various dates in the early-1980s.[22] In all cases, they note that the first emoticons were smiley faces and were intended to compensate for what users considered the dry or cold nature of text-based communication, made more sterile by the plain-text font, which could not handle font size changes, or bolding, italicizing, or underlining for emphasis. (Users active on the Internet in its early days appear to have distinguished between the slow, text-based communication of a letter and the instant communication of an email, and were concerned that the speed of the latter would negatively impact receivers' interpretation of the emotion(s) accompanying email messages.) First-generation emoticons appeared as a combination of punctuation marks: colon, dash, and right parenthesis, or: :-). These smileys became popular as a way for writers to clarify their emotional states and avoid misinterpretation by their recipients—particularly, as one recent article has argued, in an era known for irony.

By the late-1970s, Anne Fitzpatrick has suggested, the cheerfulness associated with the original yellow smiley face had been replaced by a more ironic association with the phrase "Have a nice day!" The smiley face now signaled cynicism and a sardonic, if not sarcastic, sense of humor. As a result, she argues, it was well-suited for Scott Fahlman, then a member of Carnegie Mellon's research faculty and the one she credits as the first Internet smiley user, in 1981. "After a spate of misread communications," she states, "he suggested that anyone who was posting a sarcastic comment on the bulletin board should explicitly label [their post]". He proposed the colon/dash/right parenthesis smiley face as the label for "attempted humor" and the frowning version to signal genuine unhappiness. Fitzpatrick's history is interesting in its own right, particularly as hers is the only one to mention the frowning smiley. However, her closing remarks offer the most useful argument for the continued evolution of emoticons. She notes that the early-1980s represented a transitional moment for the Internet: the users who had initially seen it as a collaborative work tool were increasingly seeing it as a tool for communication. As Internet usage evolved, emoticons would play an important supplementary role to email, website, and forum communications.[23]

As emoticons continued to evolve, they seemed to overcome initial criticism that replacing verbal articulations of complex emotional states with simplistic smiles or frowns would impoverish electronic communication and cause users' writing skills to decline. In large part, this seems due to the fact that each new generation has made emoticons more visual, hiding their punctuation mark origins behind actual, pictorial images. Scholars have hypothesized that the drive to further develop emoticons' visual qualities came from the recognition that textual emoticons "still lack[ed] the ability to communicate the details of an emotional response, such as the speaker's tone of voice or intensity of emotion."[24] While first-generation emoticons remained discrete punctuation marks, second-generation emoticons tried to address this issue by converting punctuation marks into simple, linear images: ☺. Third-generation emoticons added motion—animated emoticons that moved, bounced, or made sounds as the computer translated a series of punctuation marks into an animated image.

One of the most well-known animated emoticons of the mid-2000s was a Microsoft product that winked. Scholars have suggested that animation was intended "not only to draw the recipient's attention but also to express the sender's emotion more intensively."[25] Yet animation did not address the issues of ambiguity in tone and register that emoticons presented, in part because they could "only depict emotion in an abstract and general manner." Their only ability to indicate the sender's tone, intensity, or overall demeanor came through repeating or mixing several emoticons in a row.[26] They reflected neither the individual personality of the sender nor the overall community to which the sender and recipient belonged. As a result, scholars have proposed more sophisticated animation or even "kinetic typography" (animated text) to better address these issues. A fourth generation of emoticons has emerged, incorporating some of these proposals in diverse ways but with a common emphasis on personalization. Regardless of the technique, fourth-generation emoticons tend to better reflect the specific personal expression needs of particular communities—in this case, practicing Muslims.

In other words, the most recent development in the sphere of emoticons has been the proliferation of more personalized, community-specific emoticons—emoticons which may each be used less frequently by the overall online population than the smileys or animated smileys of previous generations, but which are likely to be used intensively within particular communities. In the case of Muslim communities online—communities in which the members are self-identified as practicing or aspirationally practicing

Muslims—the use of emoticons in general, and fourth-generation emoticons in particular, in forums and chat boards, has been both common and highly critiqued. This study now turns to several of these online forums, to examine how emoticons are used, the policies relating to emoticon usage on some of these websites, and other, clerical responses to the use of emoticons for practicing Muslims. These forums are English-language, although some also have Arabic-language partner forums and the level of English proficiency among users varies. Interestingly, physical location plays almost no part in users' discussions. While some users identify a country of location or origin via their avatar or signature, the forum threads tend to be as dislocated as they are disembodied.

ISLAM-SPECIFIC EMOTICONS

What constitutes a fourth-generation emoticon for online Muslim communities? First, many are textual, rather than visual. They reproduce Arabic-language Islamic expressions as pictorial icons, starting with the standard Muslim greeting "Salaam w `aleikum" or "peace be upon you":

They extend to phrases like "Salla Allahu `alaihi w sallam" or "May God's blessings be upon him," which is used by pious Muslims after every reference to Muhammad. Arabic-language Islamic expression emoticons also include formulaic but personal expressions like "Astaghfur Allah" or "I ask God's forgiveness." These textual emoticons range from simple, digital fonts to elaborate, calligraphic pictorial images like the one indicated in 5.1, which represents the basmala—the oft-recited phrase translating as: "In the name of God, the Most Compassionate, the Most Merciful." Emoticons like these are easily found on many sites around the Internet, with assorted text-based shortcuts that, once the website in question has activated them, allow the user to type the shortcut in Roman script and see the pictorial icon appear on his or her screen.

السلام عليكم

Figure 5.1 The standard Muslim greeting "Salaam w `aleikum" or "peace be upon you".

بسم الله الرحمن الرحيم

Figure 5.2 Arabic-language Islamic expression emoticons also include formulaic but personal expressions like "Astaghfur Allah" or "I ask God's forgiveness".

What these textual, Arabic-language Islamic expression emoticons suggest is, first, that the "icon" part of an emoticon does not need to be an image—or at least an image of a person or other animate being. Second, they suggest the continuing—or, perhaps, increasing—importance of Arabic for pious Muslims. Since the early days of Islam, there has been a broad consensus that Arabic is the only language of revelation, and hence that a translation of the Qur'an has a different and lesser status than the Qur'an itself, because the act of translation changes the words and adds the perspective of the human translator, while the Qur'an reflects the eternal and direct words of God. Yet there is no theological reason for privileging the Arabic for "I ask God's forgiveness" over the speaker's native tongue. What the appearance of these emoticons suggests, then, is the importance that Arabic plays for these self-identified practicing Muslims.

EMOTICONS AND MUSLIM ONLINE SPACES

It is difficult to obtain older user guidelines and rules of conduct for online forums and chat boards, often because they have been rewritten over time or the boards have disappeared, or sometimes because early boards did not have comprehensive, posted rules of conduct. The earliest user guidelines found in the course of this study that deal with emoticons come indirectly from a 2004 post on the Islamic Forum site. Titled "Free Islamic emoticons," the post highlighted a new set of "Islamic emoticons" that site editors had recently introduced, stating:

> [We] have made available for its users, a set of Islamic emoticons, in the form of Arabic text of the common words Muslims often use in dialogue and discussions. I've also made them available for download. They're useful for Islamic webmasters, and forum admins, to make them available for their users, by clicking on them, instead of typing the actual Arabic text. They are also useful for those who do not have Arabic keyboards available, or even for non-Muslims who would like to socialize with us, or impress someone [smiley face].
>
> The download comes with emoticons in 2 sets, one in black text, suitable for light backgrounded webpages, and one in white text, for dark colored ones. Both are transparent to suit any color scheme.
>
> To download IF Islamic emoticons, go here: http://www.gawaher.com/show.php/showtopic/2914.html
>
> For details on our Islamic emoticons and what each means in English, go here: http://forums.gawaher.com/index.php?showtopic=2820[27]

This post is interesting for several reasons. First and foremost, it casually used the term "Islamic emoticons" with no accompanying definition, suggesting that the poster assumed it would be a familiar term to readers. At the same, it neatly established the bounds of the "Islamic emoticon" category by describing them as Arabic-language, textual emoticons that displayed frequently-used pious phrases like the basmala. No pictorial emoticons were included, whether of human faces or other images. Rather than compensating for the emotional ambiguity of text-based conversations, the poster suggested that these emoticons would make life easier for posters who would otherwise have to type out these phrases in Arabic script, or who would—lacking Arabic keyboards and language packs—have to transliterate them.

This post also mentioned the technical infrastructure needed to support the use of these emoticons, noting that these emoticons needed to be downloaded and made available on individual websites for site users to be able to employ them. As a result, it focused less on promoting the availability of these emoticons on Islamic Forum and more on reaching the webmasters and forum administrators who managed other sites. This "business to business" focus reflected the fact that the decision to allow emoticons, Islamic or otherwise, rested with website owners and managers rather than individual site users. If administrators chose not to enable emoticon-recognizing code, users would type in the relevant character shortcuts without seeing them transform into images.

The link to the emoticons mentioned in this post is no longer active—a reminder of the sometimes-ephemeral nature of the Internet, especially after several years. Yet a survey of contemporary self-described "Islamic emoticons" reveals a continued emphasis on textual rather than pictorial icons, on the use of Arabic rather any other language—including English, the language used by posters promoting these emoticons—and on the use of stock pious phrases. This development appears to have emerged in tandem with discussions by religious scholars and practicing Muslims about the danger or impermissibility of pictorial emoticons, although certain pictorial emoticons remain popular and continue to circulate on self-defined Muslim sites.

Contemporary Emoticons

Emoticons identified as "Islamic emoticons" can today be found on a range of pious websites—Sunni and Shii, progressive and conservative—and with site managers based in a variety of countries. (For the purposes of this study, and because so much computer science work is done in English, only English-language sites were surveyed.) For the most part, these emoticons fall within the range of those described in the 2004 Islamic Forum post: Arabic-language, Arabic script religious phrases and expressions. However, they also include pictorial emoticons that depict self-described "Islamic images." Often these involve reworkings of generic smiley faces—depicting

Figure 5.3 "Hijab Icon".[28]

Figure 5.4 and 5.5 5.4 (left) shows a smiley saying "Ma sha' Allah", "As God has willed". 5.5 (right) shows a smiley saying "Ameen", which is close to the Christian use of "Amen".[29]

them in piously gendered ways with hijabs or caps and beards, as in the hijab icon shown in 5.3 downloaded from the Sunni site www.attari.net.

These reworked smileys often are created as .gif files, so do not work with the textual shortcuts commonly used for smileys, but users can insert them at the end of posts or in their signature lines. A third category blends both approaches, either by producing highly decorated, sometimes animated Arabic emoticons or by depicting a smiley speaking Arabic, as here:

Often, users respond to newly posted emoticons with thanks and appreciative statements ranging, as in a 2008 thread on ShiaSisters.net, from the less formal "They are so cool!," to the more pious "May Allah reward you [with] the best."[30] Yet what comes across most strongly when surveying emoticon discussions on contemporary forums and chat boards is a discourse that reflects a lack of consistency in users' knowledge of the range of emoticons available and in their satisfaction with the kinds of emoticons site administrators make available for users. The next section discusses this discourse in greater detail.

Contemporary Discourse: Users' Knowledge

This study has found evidence of Arabic-language textual emoticons being made available for users as early as 2004. Even in the late-2000s, however, users of various Islamic websites appear to have had little idea how to access these emoticons or how to produce them for their own sites. For example, in late-2008 a user posted on Talkaboutislam.com with the thread "Making Islamic emoticons—is it possible?" The user described Islamic emoticons as textual emoticons featuring phrases like "Allah akbar" and "Baraka Allah," and asked whether it would be possible to develop them for the site so they could be used alongside smileys.[31] This particular user was not a native English speaker and hence might have been less likely to know about the emoticons available on English-language websites. However, similar queries appeared throughout the late-2000s on various Islamic websites, suggesting that users were far more cognizant of the term "Islamic emoticons" than of their availability on similar websites or even of their feasibility. Yet what comes through clearly in users' posts is their desire for Islamic emoticons and, often, their discomfort with smileys and other generic, non-Islamic ones.

Contemporary Discourse: Users and Administrators

Discussions about Islamic emoticons between users and site administrators take place against what is often described as a "crisis of authority" in the contemporary Muslim world. The historic sources of knowledge on what it meant to be Muslim and to live a Muslim life—the `ulama—have seen their prestige erode since the early-1800s, due partly to some `ulama's collaborationist behavior during the colonial period, partly to the cooption of historically independent educational institutes like al-Azhar by modern states, and most notably to the vast increase in education and literacy rates in the Muslim world. This last process transformed education from a limited privilege enjoyed almost exclusively by the `ulama and other elites, to a widespread characteristic of anyone with a secondary school or college education. As a result, education in general and literacy in Arabic in particular have become understood as the qualifications necessary for a pious Muslim to read and interpret the Qur'an and ancillary texts for him or herself, regardless of whether that education includes training in `ulum al-din (the "sciences of religion") or whether it focused on, for example, engineering.

Yet what numerous Muslim websites suggest is a deeply felt desire for religious guidance and some kind of authoritative interpretation of what constitutes a Muslim life. With respect to the issue of whether emoticons can be included in this life—at all or of certain kinds—this is most evident in the requests for fatwa on the use of emoticons and in the posts by pious Muslims about the same issue. The first involves `ulama and suggests more traditional notions of authority in Islam; the second involves ordinary educated Muslims and suggests more contemporary, decentralized, and de-hierarchized notions of authority. Both reflect an ongoing desire among Muslims for authoritative pronouncements, even if the source of authority is understood differently.

For example, one fatwa reprinted on the British website Dawah2Islam ("Call to Islam") comes from Sheikh Zayd al-Madhkalee, a Saudi sheikh born in the late-1930s. In response to a question about the use of smileys in all-male chat forums, al-Madhkalee answered: "There is no basis in the shari`a for these smileys." Explaining that "embodied" figures could be used only in case of necessity, he stated that chatting was not a necessity and reiterated that smileys lack a basis in Islam. "If these persons are in need of [communicating with] each other, than they can talk in a different way," he concluded, adding that emoticons had no place in the "Manhaj of Ahl al-Sunnah wa al-Jama`a."[32] (Ahl al-Sunnah wa al-Jama`a—"the people of the (Prophet's) Path and the community"—is a term Sunnis use to describe themselves, while "manhaj"—"methodology"—is a term used more frequently by Salafi Muslims, who have a more austere interpretation of Islam than most other Sunnis.)

What is interesting about this fatwa is the narrow range of emoticons described, as well as the specifically Wahhabi and Salafi anxieties about pictorial representation. Both questioner and sheikh refer only to smileys—to the generic pictorial emoticons that communicate happiness, sadness, etc. They do not address the phenomena either of Islamic emoticons in general or of textual emoticons in particular. While on the sheikh's part this might simply have reflected his age and lack of electronic experience, for the questioner it also suggests the scope of his concerns and, potentially, his interests. It might be that he considers textual emoticons uncontroversial and hence feels no need to ask for guidance about their use. It might be that as a native Arabic speaker he chats in Arabic script already and hence is unaware of such textual shortcuts. Or it might reflect a broader concern with appropriate chat behavior outside the bounds of Islamic websites and chat forums—a concern that the sheikh appears to share, breaking his answer into one part on emoticons and one part on the mode by which people communicate. More broadly, this fatwa suggests an ongoing desire for authoritative guidance as well as the importance of establishing the particular context envisioned by questioner and sheikh.

Islamic websites and associated chat forums, which are managed by webmasters and forum moderators, might serve as an example of the more contemporary, decentralized understanding of authority in Islam. Here, one crucial point to remember is that decentralization also often facilitates multiplicity. For example, ahlalhadeeth.com, a heavily used conservative Sunni chat forum site that describes itself as the "meeting place of students of knowledge," displays in its forum posts and moderator statements a disapproval of emoticons not reflected in the site's usage policies, which permit a range of generic smileys. The site's FAQ section defines smileys as "small graphical images that can be used to convey an emotion or feeling" and lists the shortcuts, graphics, and meanings of those activated on the site, including "Smile," "Stick Out Tongue," and "Roll Eyes (Sarcastic)."[33] Yet user posts about smileys and emoticons more broadly appear almost all hostile to their use. For example, a long discussion thread begun in July 2009 focused on the impermissibility of women in particular using emoticons in mixed-gender discussions—a position generally endorsed by posters, although a number also agreed that men should do the same.[34] The discussion was amplified by posters' comments that one of the site moderators, Ayman bin Khaled, had the previous year posted an extensive essay about the use of emoticons, which used hadith and other scholars' conclusions to argue that emoticons were forbidden by analogy to the prohibition on imitating God's power to create life. While bin Khaled did not address textual emoticons, his essay was unequivocal in its condemnation of emoticons and their users.[35]

A threaded discussion in 2010 returned to the issue of emoticons on the site, addressing bin Khaled with the question: "Smilies are haram so why do you have them?" He replied by explaining that the site's managers disagreed: "Some said it is haram while others said it does not reach the level of haram." Further, he noted, those with the latter view were in charge of site code. However, he added, "I am sure they are considering removing them and replac[ing] them with Islamic symbols such as subhan Allah"—in other words, replacing the generic smileys with textual emoticons using pious Arabic phrases. He closed by suggesting that the woman who started the thread ask site administrators to request textual emoticons, noting that the Arabic-language version of the website had replaced pictorial emoticons with textual ones "due to the many requests to remove them."[36] Interestingly, he ended his post with "Wallahu A'lam," "And God is the one who knows," the classical formulation for ending a fatwa or legal ruling.

The Ahlahadeeth threads suggest two points. First, that users and site administrators hold a range of views regarding the use of pictorial emoticons. In this case, users on the whole appear to have been more conservative than administrators in their interpretation of the admissibility of pictorial icons either at all or in mixed-gender chats. The discussions appear to make no distinction between generic pictorial emoticons and those that have been reworked in a more self-consciously Islamic vein, like the muhajaba or bearded smileys mentioned earlier. (Users appear much more likely to distinguish between these and generic smileys, while site administrators often list generic and Islamic emoticon shortcuts together, with no differentiation, as on Sunniport.com.)[37] Second, the threads allude to a general consensus that textual emoticons, or at least those focused around Islam, are non-controversial and generally a positive addition to online conversations. Finally, the threads focus on users' opinions rather than clerics'—in no threaded discussion, for example, did a user ask a cleric for a fatwa or other legal opinion regarding emoticons—demonstrating the flattening of authority characteristic of the contemporary period.

CONCLUSIONS

What lies behind the desire for Islamic emoticons? This study has focused primarily on the phenomenon of Islamic emoticons: placing their development within the historical context of emoticon development more generally, and charting the ways in which users and administrators on self-identified Islamic websites request, offer, and discuss Islamic and other emoticons. Yet these discussions focus far less on why users might want emoticons or what role they might play in online conversations, and far more on issues of appropriateness with respect to mixed-gender conversations and the use of images.

On one level, it appears that users on Islamic websites want emoticons for the reasons that others have: to supplement or compensate for the flatness and ambiguity of textual communication. The requests for textual emoticons depicting Arabic-language, Arabic script phrases both build on this issue and reflect the global nature of the contemporary Muslim community, the vast majority of whose members do not speak Arabic. As a result, textual emoticons help by replacing the challenge of spelling words in Arabic or typing in Arabic script. Beyond issues of compensation and communication, however, what this conclusion suggests is that Islamic emoticons, textual and visual, help provide their users with religious credibility. They offer a distinctive, supra-textual demonstration of users' commitment to a Muslim identity and personal piety. At the same time, they also help reinforce users' sense of community—particularly the textual emoticons, whose Arabic-language phrases are much more likely to be understood by practicing Muslims than others. Yet in their iconic use of Arabic, they also highlight the globality of the contemporary Muslim community, for whom Arabic fluency is rare. Arabic, as the language of the Qur'an and of prayer, has a value that goes beyond Muslims' ability to speak it—it has a visual value that signals membership in a believing community. After all, Arabic-speaking Muslims would be able to type or transliterate Arabic phrases and communicate in Arabic—perhaps on the Arabic-language forums that often exist in partnership with the English-language ones examined here.

This use of Arabic—as iconic, textual emoticons—might be understood as both compensatory and productive. It compensates for users' lack of or limited facility with Arabic and/or typing in Arabic, while indicating the continuing importance of Arabic and the Arabic script for members of these online communities. (The forums showed no examples of Arabic transliterated, as is often the case in contemporary Arabic-language text messages, Facebook posts, and so on.) Yet it might also be understood as productive—as helping instill in users an emotional connection to pious expressions in the Arabic language and the Arabic script. In this way, Arabic-language textual emoticons complicate and enrich the arguments of scholars like Saba Mahmood and Salwa Ismail. Mahmood and Ismail have worked in complementary ways to critique and reconfigure notions of the public sphere most famously articulated by Jurgen Habermas.[38]

Mahmood and Ismail have each argued that the exclusion of religion and emotion from the Habermasian public sphere has served to limit the participation of numerous 'others,' including people of faith, ethnic minorities, people of lower socioeconomic status, homosexuals, and women. Both scholars use their work with pious Sunni Muslim Egyptian women to critique normative discussions of Islamism as an unwelcome and backwards intrusion on the public sphere. As Ismail notes, "The view of Islamism as anti-modern rests on the assumption that modernization is associated with secularization and the retreat of religion from the public sphere. Islamism thus appears as an expression of an anti-modern strand that, for some, is inherent in the religion."[39] Instead, they look at the cultivation of a particular kind of self-hood—which Mahmood describes as part of a "piety movement" and Ismail terms a "Muslim selving." For both, physicality plays an important role in these processes—whether in cultivating a particular sensibility of pious tears or training oneself into the discipline of daily prayers.[40] Yet when it comes to emoticons, the work being done—the cultivation of self and sensibility—is stripped of its physicality. Instead, with emoticons, the work of producing and sustaining a particular affect shifts from the physical to the visual realm—which lays out a new set of questions to address, as well as adding a new arena through which to contemplate notions of the public sphere (or spheres).

What these two aspects—personal credibility and community feeling—suggest is that for users Islamic emoticons might do much more than compensate for the flatness of text-based communication. As a result, this study suggests, further research is needed that takes seriously the work that Islamic emoticons do, and the ways in which they might do it. The coalescence of a common understanding of what "Islamic emoticons" encompass, as well as the disparate views on the appropriateness of other emoticons on Islamic websites, suggest that users employ emoticons for both personal expression and to affiliate with a particular community. Yet further research may discover that these emoticons do more: that, along the lines of the research done by scholars like Saba Mahmood,[41] they facilitate users' pious self-cultivation, providing a visual reminder of the particular sensibilities endorsed within these online community spheres.

NOTES

1. An early version of this work was presented at the International Conference on Digital Religion, held at the University of Colorado at Boulder in January 2012. The author would like to thank participants at that conference for their helpful comments, as well as the editors of this volume for the opportunity to further explore this important issue.
2. For a sample of their older work, see Bunt,G. (2003) *Islam in the Digital Age: E-Jihad, Online Fatwas and Cyber Islamic Environments*. New York: Macmillan; and Anderson, J. (2003) New Media, New Publics: Reconfiguring the Public Sphere of Islam, *Social Research*, 70(3, Fall), 887–906.
3. Marcotte, R. (2010) Gender and Sexuality Online on Australian Muslim Forums. *Contemporary Islam*, 4(1), 117–138.
4. Mahmood, S. (2004) *The Politics of Piety: The Islamic Revival and the Feminist Subject*. Princeton, NJ: Princeton University Press.

5. See, for example, Rezabek, L. and Cochenour, J. (1994) Emoticons: Visual Cues for Computer-Mediated Communication, in *Imagery and Visual Literacy: Selected Readings from the Annual Conference of the International Visual Literacy Association* (Tempe, Arizona, October 12–16); as well as their later article (1998) Visual Cues in Computer-Mediated Communication: Supplementing Text with Emoticons, *Journal of Visual Literacy*, 18(2, Autumn), 201–215. Another early study was conducted by Rivera, K., Cooke, N.J. and Bauhs, J.A. (1999) The Effects of Emotional Icons on Remote Communication. [Interactive poster]. *Proceedings: CHI '96: Conference Companion on Human Factors in Computing Systems: Common Ground*.
6. Dresner, E. and Herring, S. (2010) Functions of the Nonverbal in CMC: Emoticons and Illocutionary Force. *Communication Theory*, 20(3, August), 249–268.
7. Derks, D., Bos, A.E.R. and von Grumbkow, J. (2008) Emoticons and Online Message Interpretation. *Social Science Computer Review*, 26(3, Fall), 379–388.
8. Derks, D., Bos, A.E.R. and von Grumbkow, J. (2008) Emoticons in Computer-Mediated Communication: Social Motives and Social Context. *CyberPsychology and Behavior*, 11(1, February), 99–101.
9. Tung, F.-W. and Deng, Y.-S. (2007). Increasing Presence of Social Actors in E-Learning Environments: Effects of Dynamic and Static Emoticons on Children. *Displays*, 4–5(December), 174–180.
10. Lo, S.-K. (2008) Rapid Communication: The Nonverbal Communication Functions of Emoticons in Computer-Mediated Communication. *CyberPsychology and Behavior*, 11(5, October), 595–597.
11. McDougal, B.R., Carpenter, E.D. and Mayhorn, C.B. (2011) Emoticons: What Does This One Mean? *Proceedings of the Human Factors and Ergonomics Society Annual Meeting*, 55(1, September), 1948–1951.
12. Anderson, J.W. and Eickelman, D. (Eds.). (1999) *New Media in the Muslim World* (rev. ed.). Bloomington, IN: Indiana University Press.
13. Lerner, M.Y. (2010) Connecting the Actual with the Virtual: The Internet and Social Movement Theory in the Muslim World: The Cases of Iran and Egypt. *Journal of Muslim Minority Affairs*, 30(4), 555–574.
14. Hashemi-Najafabadi, S.A. Has the Information Revolution in Muslim Societies Created New Publics? *Muslim World Journal of Human Rights*, 7(1), Article 4.
15. See, for example, Howard, P.N. (2010) *The Digital Origins of Dictatorship and Democracy: Information Technology and Political Islam*. Oxford: Oxford University Press.
16. Lim, F.K.G. (Ed.). (2009) *Mediating Piety: Technology and Religion in Contemporary Asia*. Leiden: Brill.
17. See, for example, Ho, S.S., Lee, W. and Hameed, S.S. (2008) Muslim Surfers on the Internet: Using the Theory of Planned Behavior to Examine the Factors Influencing Engagement in Online Religious Activities. *New Media and Society*, 10(1, February), 93–113.
18. See, for example, Marcotte, Gender and Sexuality Online on Australian Muslim Forums.
19. Scholz, J., Selge, T., Stille, M. and Zimmermann, J. (2008) Some Remarks on the Construction of Religious Authority in Islamic Podcasts. *Die Welt des Islams*, 48(3–4), 457–509.
20. Warf, B. (2010) Islam Meets Cyberspace: Geographies of the Muslim Internet. *The Arab World Geographer*, 13(3–4; Fall–Winter), 217–233.
21. Bunt, G.R. (2009) *iMuslims: Rewiring the House of Islam*. Chapel Hill, NC: University of North Carolina Press.

96 Andrea L. Stanton

22. For the 1979 date, see Preece, J., Maloney-Krichmar, D. and Abras, C. History of Online Communities. In K. Christensen and D. Levinson (Eds.) (2003), *Encyclopedia of Community: From Village to Virtual World* (pp. 1023–1027). Thousand Oaks, CA: Sage Publications. It should be noted here that this innovation did not represent the invention of the smiley face: the now-iconic yellow circle with two black dots for eyes and a simple curved mouth was designed by the American graphic designer Harvey Ball as part of a campaign by the State Mutual Insurance Company to improve customer service and employee morale.
23. Fitzpatrick, A. (2003) On the Origins of :-). *IEEE Annals of the History of Computing*, July–September, 82–83.
24. Lee, J., Jun, S., Forlizzi, J. and Hudson, S.E. (2006) Using Kinetic Typography to Convey Emotion in Text-Based Interpersonal Communication. *Proceedings of the 6th Conference on Designing Interactive Systems* (pp. 1–10). New York: ACM.
25. Lee, J., Jun, S., Forlizzi, J. and Hudson, S.E. (2006) Using Kinetic Typography to Convey Emotion in Text-Based Interpersonal Communication.
26. Lee, J., Jun, S., Forlizzi, J. and Hudson, S.E. (2006) Using Kinetic Typography to Convey Emotion in Text-Based Interpersonal Communication: 2.
27. Source: http://www.gawaher.com/index.php?showtopic=2915, retrieved November 6, 2011. The links in the quoted text were no longer active. This image appears courtesy of http://gawaher.com, and can be found at: http://www.gawaher.com/topic/2820-arabic-shortcuts-smileys/
28. http://www.attari.net/?Islam:Islamic_MSN_Emoticons, retrieved December 3, 2011. The site suggests that users download these pictorial emoticons and create shortcuts for them in the "custom emoticons" section of Microsoft's instant messaging window. This particular emoticon came from http://www.emofaces.com/emoticons/muslim-emoticon; license to reprint this emoticon was obtained March 18, 2013.
29. http://www.islamictorrents.net/smilies.php, retrieved September 30, 2011. These smileys were added in 2006 after users requested more Islamic emoticons—with at least one user providing a list of the phrases he wished to see made available.
30. http://shiasisters.net/forum/viewtopic.php?t=128, retrieved December 11, 2011.
31. http://talkaboutislam.com/forums/index.php?topic=1000564.0, October 23, 2008, retrieved December 1, 2011.
32. http://www.dawah2islam.co.uk/apps/blog/show/4551860-fatwa-on-chat-smileys, posted August 18, 2010, retrieved December 5, 2011.
33. "Smilies," http://www.ahlalhdeeth.com/vbe/misc.php?s=ba909d4caeeba1c31b8298add29a8296&do=showsmilies, retrieved December 5, 2011. This page also notes that readers can disable smileys in their posts but does not suggest why, other than preserving program code, they might want to do so.
34. Sisters Using Emoticons on Islamic Forums, http://www.ahlalhdeeth.com/vbe/showthread.php?t=5737&highlight=emoticons, retrieved December 5, 2011.
35. bin Khaled, A. (2008, February 19) Emoticons. Retrieved December 5, 2011 from http://ahlalhdeeth.com/vbe/showthread.php?t=1625
36. Smilies Are Haram So Why Do You Have Them?, Retrieved December 5, 2011 from http://www.ahlalhdeeth.com/vbe/showthread.php?t=8287&highlight=emoticons. Similar threads exist on other sites, like a 2006 one on Turntoislam.com, retrieved December 1, 2011 from http://www.turntoislam.com/forum/showthread.php?t=5329&highlight=smileys, and one the previous

year on Understandingislam.com, retrieved December 1, 2011 from http://forums.understanding-islam.com/archive/index.php/t-1269.html?s=c2161656c809bc34ca8678b4cbd37802
37. http://www.sunniport.com/masabih/misc.php?do=showsmilies, retrieved December 14, 2011.
38. Mahmood, *The Politics of Piety*; and Ismail, S. (2007) Islamism, Re-Islamization and the Fashioning of Muslim Selves: Refiguring the Public Sphere. *Muslim World Journal of Human Rights*, 4(1), 1–21.
39. Ismail, S. (2006) *Rethinking Islamist Politics: Culture, the State, and Islamism* (2nd ed.). London: I.B. Tauris: 3.
40. Mahmood, S. (2001) Spontaneity and the Conventionality of Ritual: Disciplines of Salat. *American Ethnologist*, 28(4), 827–853.
41. Mahmood, The Politics of Piety.

REFERENCES

Anderson, J. (2003, Fall) New Media, New Publics: Reconfiguring the Public Sphere of Islam. *Social Research*, 70(3), 887–906.
Anderson, J.W. and Eickelman, D. (Eds.). (1999/2003) *New Media in the Muslim World* (rev. ed.). Bloomington, IN: Indiana University Press.
Bunt, G.R. (2009) *iMuslims: Rewiring the House of Islam*. Chapel Hill, NC: University of North Carolina Press.
Bunt, G. (2003) *Islam in the Digital Age: E-Jihad, Online Fatwas and Cyber Islamic Environments*. New York: Macmillan.
Derks, D., Bos, A.E.R. and von Grumbkow, J. (2008, February) Emoticons in Computer-Mediated Communication: Social Motives and Social Context. *CyberPsychology and Behavior*, 11(1), 99–101.
Dresner, E. and Herring, S. (2010, August) Functions of the Nonverbal in CMC: Emoticons and Illocutionary Force. *Communication Theory*, 20(3), 249–268.
Fitzpatrick, A. (2003, July–September) On the Origins of :-). *IEEE Annals of the History of Computing*, 82–83.
Hashemi-Najafabadi, S.A. (2010) Has the Information Revolution in Muslim Societies Created New Publics? *Muslim World Journal of Human Rights*, 7(1), Article 4.
Ho, S.S., Lee, W. and Hameed, S.S. (2008, February) Muslim Surfers on the Internet: Using the Theory of Planned Behavior to Examine the Factors Influencing Engagement in Online Religious Activities. *New Media and Society*, 10(1), 93–113.
Howard, P.N. (2010) *The Digital Origins of Dictatorship and Democracy: Information Technology and Political Islam*. Oxford: Oxford University Press.
Ismail, S. (2006) *Rethinking Islamist Politics: Culture, the State, and Islamism* (2nd ed.). London: I.B. Tauris.
Ismail, S. (2007) Islamism, Re-Islamization and the Fashioning of Muslim Selves: Refiguring the Public Sphere. *Muslim World Journal of Human Rights*, 4(1), 1–21.
Lee, J., Jun, S., Forlizzi, J. and Hudson, S.E. (2006) Using Kinetic Typography to Convey Emotion in Text-Based Interpersonal Communication. *Proceedings of the 6th Conference on Designing Interactive Systems*, 1–10. New York: ACM.
Lerner, M.Y. (2010) Connecting the Actual With the Virtual: The Internet and Social Movement Theory in the Muslim World: The Cases of Iran and Egypt. *Journal of Muslim Minority Affairs*, 30(4), 555–574.
Lim, F.K.G. (Ed.). (2009) *Mediating Piety: Technology and Religion in Contemporary Asia*. Leiden: Brill.

Lo, S.-K. (2008, October) Rapid Communication: The Nonverbal Communication Functions of Emoticons in Computer-Mediated Communication. *CyberPsychology and Behavior*, 11(5), 595–597.

Mahmood, S. (2001) Spontaneity and the Conventionality of Ritual: Disciplines of Salat. *American Ethnologist*, 28(4), 827–853.

Mahmood, S. (2004) *The Politics of Piety: The Islamic Revival and the Feminist Subject*. Princeton, NJ: Princeton University Press.

Marcotte, R. (2010) Gender and Sexuality Online on Australian Muslim Forums. *Contemporary Islam*, 4(1), 117–138.

McDougal, B.R., Carpenter, E.D. and Mayhorn, C.B. (2011, September) Emoticons: What Does This One Mean? *Proceedings of the Human Factors and Ergonomics Society Annual Meeting*, 55(1), 1948–1951.

Preece, J., Maloney-Krichmar, D. and Abras, C. (2003) History of Online Communities. In K. Christensen and D. Levinson (Eds.), *Encyclopedia of Community: From Village to Virtual World* (1023–1027). Thousand Oaks, CA: Sage Publications.

Rezabek, L. and Cochenour, J. (1994) Emoticons: Visual Cues for Computer-Mediated Communication. *Imagery and Visual Literacy: Selected Readings from the Annual Conference of the International Visual Literacy Association*, Tempe, Arizona, October 12–16.

Rezabek, L. and Cochenour, J. (1998, Autumn) Visual Cues in Computer-Mediated Communication: Supplementing Text with Emoticons. *Journal of Visual Literacy*, 18(2), 201–215.

Rivera, K., Cooke, N.J. and Bauhs, J.A. (1996) The Effects of Emotional Icons on Remote Communication. [Interactive poster]. *Proceedings: CHI '96: Conference Companion on Human Factors in Computing Systems: Common Ground*, April 13-18, 1996, Vancouver. Published by the Association for Computing Machinery, New York, 1999.

Scholz, J., Selge, T., Stille, M. and Zimmermann, J. (2008) Some Remarks on the Construction of Religious Authority in Islamic Podcasts. *Die Welt des Islams*, 48(3–4), 457–509.

Tung, F.-W. and Deng, Y.-S. (2007, December) Increasing Presence of Social Actors in E-Learning Environments: Effects of Dynamic and Static Emoticons on Children. *Displays*, 4–5, 174–180.

Warf, B. (2010, Fall–Winter) Islam Meets Cyberspace: Geographies of the Muslim Internet. *The Arab World Geographer*, 13(3–4), 217–233.

6 Emotional Socialization on a Swedish Internet Dating Site
The Search and Hope For Happiness

Henrik Fürst

INTRODUCTION

The Internet is an important place for finding a partner in Sweden today.[1] But finding a partner through Internet dating is not something that happens instantaneously. For those looking and longing for a partner, the search is a process that extends over time. In this study I focus on the hope and search for a partner on one particular Swedish Internet dating site. The participants' emotional experiences of Internet dating practiced over time are studied through email interviews with participants at one point in time. Hence, Internet daters are defined as the responding participants active on the Internet dating site at the moment of my data collection.

Previous research about Internet dating has focused on deception and trust (Hardey 2004; Lawson and Leck 2006); the use of Internet dating in order to overcome a life crisis (Couch and Liamputtong 2008; Lawson and Leck 2006); the use of Internet dating sites to find sex partners (Couch and Liamputtong 2008; Daneback et al. 2007); self-presentations and authenticity (Hancock et al. 2007; Hardey 2004; Whitty 2007); Internet dating as immaterial labor (Arvidsson 2006); and Internet dating as an example of where emotional relationships have become economically interesting (Illouz 2007). No earlier studies have explicitly focused on the problem formulated in this text: (1) the emotional experiences amongst the members of an Internet dating site over time, and (2) the related claim as to the importance of emotions in reproducing Internet dating-related activities.

The core idea of this chapter is that the participants have an emotional commitment of hope to the activity of searching for a partner through Internet dating, which is negotiated over time. This emotional commitment reproduces Internet dating activities. It is nurtured and commercialized by Internet dating companies (see also Arvidsson 2006; Illouz 2007).

This study is based on a remodeled version of the social worlds/arenas framework. This framework derives from interactionist grounded theory (Clarke 2005). I have kept the interactionist and pragmatist roots of the framework, but I have remodeled it in order to establish research problems that emphasize the importance of emotions in Internet dating practiced over time. The theory/method package is associated with a number of sensitizing concepts, which serve as pathfinders for discovering patterns in the material and creating new sensitizing concepts from the material (Clarke 2005).

THEORY/METHOD PACKAGE

Material

In order to access data about Internet dating experiences, I contacted an Internet dating company. In late-2009, a bulk email was sent out by an administrator of the Internet dating site to 600 email addressees registered on the website. The email contained information about the study and questions related to the theme of this research. Having received 68 responses to the initial email interview, I carried out follow-up email interviews with 15 of the interviewees. Whereas the initial email responses were short, the follow-up email responses were more elaborate on matters such as emotions and activities. However, due to this self-selection process, participants could choose to write a response or refrain from it, and hence not all Internet dating activities are covered, but rather, the most legitimate uses of Internet dating—which the participants call "being serious." Nonetheless, other "illegitimate" positions are talked about. Furthermore, in order to study the Internet dating companies' role in generating Internet dating, e.g., by creating phantasmagorias, I also analyzed the front page of the Internet dating site.

Coding Procedure

The material was coded according to a basic grounded theory approach. Concepts were formed through labels given to elements in the material and by comparing and relating them to emerging concepts (Glaser 1978: 62). In this manner, narrations about experienced emotions in Internet dating were coded and conceptualized.

A Social Worlds/Arenas Framework

Social worlds are a collective commitment to some sort of activity (Strauss 1978, 1982, 1984, 1993; Becker 1982; Clarke 1991, 2005). This is what people actually do together by sharing a commitment to some sort of activity that defines that particular social world. In this respect, it is possible to analyze Internet dating activities as the shared actor-defined reality of Internet dating, i.e., the Internet dating world. Related social worlds meet on social arenas where they struggle over shared resources (Strauss 1978). Internet dating is part of a social arena of dating. The social worlds/arenas framework makes it possible to transcend media-centrism. This perspective is particularly appropriate when it comes to Internet dating, which not only occurs on the Internet but involves other sites for action as well.

The people I interviewed belong, and are to various degrees committed, to the Internet dating world. The idea is that we belong to various social worlds in our lives (Luckmann 1978; Strauss 1982), and the Internet dating world can be one of them. As social beings, we are part of social worlds and we do things together in similar ways and on certain associated sites for action. We also use associated technologies for carrying out the activities of the social world (Strauss 1978). The Internet dating site may be seen as a particular site for action and as containing various technologies for carrying out Internet dating activities.

The Use of Sensitizing Concepts

Using the social worlds/arenas framework means utilizing sensitizing concepts. Sensitizing concepts function as guiding metaphors, but do not determine the outcome of the coding and analysis (Blumer 1969: 147–148). They are related to a prior picture or scheme of the field of inquiry (Blumer 1969: 24–25). I argue that this approach to data can be called a "sensitizing analytical strategy," i.e., a strategy which allows to automatically think "interactionally, temporally, processually, and structurally" (Strauss 1993: 67–68; see also Clarke 2005). For example, one sensitizing concept that has been used is "primary activity" in a social world, i.e., an activity that has come to define that particular social world (see Strauss 1978). The search and hope for a heterosexual monogamous relationship with a romantic future partner has been identified as a prominent primary activity. This search constitutes a defining core social process in practicing Internet dating on this particular Internet dating site. Furthermore, the coding procedure allowed me to construct new sensitizing concepts during my analysis.

Analyzing and Explicating Situational Influences, Positions in the Data and Social Worlds/Arenas

Sensitizing concepts should also be part of an analysis of the broader situation of action (Clarke 2005). Situational analysis involves three analytical strategies. Mapping the situation of action means, for example, to map out economic influences, symbolic manifestations, individual and collective actors, non-human and human actors, temporal aspects, etc. Further mapping involves laying out salient or implicated positions taken and not taken by the actors in the material in respect to some emerging category. Also, situational analysis involves mapping out social worlds and social arenas with the help of the sensitizing concepts. The social worlds/arenas produce discourses about how to proceed with the activity of a particular social world in its social arena. The concept of discourse is understood as actualizing the issue of power in constructions of meaning and definition of a situation (Clarke 2005: 149). I will treat the term phantasmagoria as a discourse relating to imagination and fantasies about sought-for future life situations, as contrasted to the current life situation.

Constructed Sensitizing Concepts Relating to Emotions in Internet Dating

I analyzed emotional experiences as the narrative accounts of emotions representing felt experiences. More specifically, the focus is on the meaning of the recollected emotions in situations of Internet dating.

One theory of emotion understands emotions as arising from disruption of habitual actions in interaction. The resulting felt experiences constitute a basis for subsequent actions. Put differently, emotions arise through the disruption of one's initial habitual tendency to act (Mead 1982: 40–43; Mead 2001: 3–8; see also Denzin 1983, 1984; Engdahl 2005). For example, a person caught in a violent situation with no way of escaping will experience fear, because the initial tendency to walk away from the threatening situation has been disrupted (see Mead 1910: 178–179; Mead 1982: 40–43). The role of emotion is to direct the actor, in the immediate situation, toward the completion of the disrupted act (see also von Wright 2000: 98). In the previous example, completion of the disrupted act is to flee from the threatening situation. An emotion is thus an early predisposition to the completion of a disrupted act in the future (Mead 1982: 40–43; Mead 2001: 3–8).

Hochschild (1983) writes about the capacity in humans to reflectively control their emotional displays. Contemporary corporations sometimes pressure employees to give certain emotional displays. Hochschild argues that employees perform emotional labor in that context, displaying emotions that are preferred by the employer. An example of such display is the smiling of flight attendants (Hochschild 1983). Emotional labor, or emotion management, is about changing or suppressing one's actual emotional experiences and displaying expected emotions which fit the situation, often as defined by someone else (Hochschild 1983: 8, 17–18). An adjacent concept is feeling rules in relation to a situation. Feeling rules guide actions and determine which emotions should be felt, for how long, in what circumstances, and in which intensity (Hochschild 1983: 56). Furthermore, whereas feeling rules structure actions, they are realized by people's interaction within a situation structured by such feeling rules (Strauss 1993: 134).

Instead of talking about feeling rules, such as those controlled by company standards, I will introduce my sensitizing concept *emotional ordering*. I argue that participants in a social world can align themselves with or deviate from the emotion(s) that create a commitment to a primary activity that forms the core social process of the social world. As the participants align with the emotion(s), the core social process is energized. An emotional ordering is an underlying pattern of emotions ordering the social world and energizing its social processes. This emotional ordering produces shared activities that keep the social world together. There are possibly competing emotional orderings, but some emotional orderings become more legitimate than others. I also argue that companies can be part in shaping, tuning, using, and evoking an emotional ordering of the social world. They can introduce products and situations to generate certain emotions, and thus activities, which are then made profitable (see also Illouz 2007). The participants act as prosumers (Ritzer and Jurgenson 2010): they produce content on the site and consume each other's content.

I also developed the sensitizing concepts *emotional socialization* and *emotional careers*. Emotional socialization is a trajectory. It is the negotiation of commitment to an activity, such as Internet dating, pursued over time, where participants come to deviate from and adhere to the emotional ordering of the social world. In this case, the trajectory over time forms an emotional attitude where one is in control of one's emotions. The pathways of emotional socialization, which appear to be common to many participants, are what I call the *emotional careers*. In Internet dating, there is an emotional career of hope.

FINDINGS

In this section I will discuss ways of practicing Internet dating and the Internet dating company's role in Internet dating. I will focus on the emotional experiences of practicing Internet dating over time, and the Internet dating company's influence on the process. The section ends with a discussion about the emotional career and emotional socialization in Internet dating at this particular Internet dating site.

Practicing Internet Dating in the Dating Arena

In Internet dating, a primary activity is to search for a future romantic partner. The participants mention other worlds in the dating arena where they can meet a partner: dating acquaintances, night life, meeting someone through work, and random meetings. This is exemplified by a participant who argues that Internet dating stands out as the most promising strategy for her (all quotes that follow are my translation):

> I came in contact with internet dating when I was single a couple of years ago. I was not up for going to night clubs and I had no dateable men "left" among my acquaintances.
>
> (Female, 30 years old)

In the Internet dating world, the participants negotiate the legitimacy of their activity of meeting people, as assumed to be illegitimate in other worlds, by, for example, claiming that they are doing "serious business": they are looking for a future partner. This decreases the level of embarrassment in practicing Internet dating, to which each individual might otherwise be subjected. Another associated strategy is to downgrade the legitimacy of the other worlds in the dating arena.

This primary activity in Internet dating is based on the core social process of being "in search for a future partner/relationship," as will be exemplified later in the analysis of the front page of the Internet dating site. On the Internet dating site, people thus have a common orientation toward dating. Because of this idea, people are evaluating each other in Internet dating. This makes the activities centered on figuring out if the other person is one's future partner. The structure of the Internet dating and the shared activity of dating create a sense of certainty of what the situation is all about. Dating in other social worlds may be a more uncertain business, since it is not necessarily a primary activity of those worlds. Also, by participating in the common activities of the Internet dating site, a specific version of dating and Internet dating is promoted and learned by the participants. As the following quote indicates, it is plausible that other Internet dating sites produce other forms of relationship and desired results of dating.

> At [this Internet dating site] I look for a girlfriend. At [another Internet dating site] I look for friends and/or a girlfriend. At yet [another Internet dating site] I look for someone to have sex with or a girlfriend.
> (Male, 43 years old)

Other participants say they use the particular Internet dating site on which this chapter focuses to find friendship. This primary activity is often presented as a side activity to looking for a future romantic partner. But as expressed by a 43-year-old man: "many singles totally ignore things such as making new friends. It becomes 'all or nothing.'"

People also engage in Internet dating in order to be approved, to gain a feeling of self-worth based on others' evaluation of oneself as desirable on a market of presumptive partners. This forms yet another, more covert, primary activity on the Internet dating site. It is sometimes interwoven with other people looking for a future partner, as obtaining approval means being evaluated as desirable by those other people. This activity can also occur while looking for a romantic future partner. The female participant quoted next acknowledges that people look for a "steady relationship," whereas she does something that might be valid for her and not for other participants.

> I look for friends and flirts rather than a steady relationship. In other words, I look for sporadic meetings, and if the chemistry is right, I could have a sexual relationship with someone. I personally do not look for a steady partner, and my goal is not to meet someone to have sex with either. The best thing about being on an internet dating site is

that you get extreme amounts of approval, and this is probably what I am primarily looking for.

(Female, 43 years old)

Energizing Internet Dating Through Hope of Happiness in Phantasmagorias

> [T]o be honest, doing internet dating is a very efficient way to avoid the feeling that you're lacking someone to have by your side. It feels very good when you do not feel alone after work, because you get a flirt from someone or you just get to talk to someone and you hope the right person enters your life. Actually, you get a bit addicted to this feeling. I have never met someone with whom it feels as though it might be the one. I wonder if you ever do this, but internet dating gives you hope that you might one day meet someone who you would never have met otherwise.
>
> (Female, 25 years old)

Here again, feeling approved takes a prominent place in the narrative and is talked about as something you get addicted to. In this case it is not an end in itself, but the feeling of approval brings about hope of finding the "right one." For those practicing Internet dating, hope as an emotion is experienced in the present, as their current life situation meets with resistance in relation to the idea of a future partner. Hope orients the participants of Internet dating toward completion of the Internet dating in the future. Hope forms an ongoing trajectory with an end point in the future. In the future lies happiness with a future partner, and thus the intentional character of experiencing hope is directed toward yet another emotional state in the future. Some of the participants equate happiness with falling in love, as the following participant seems to do.

> [I started Internet dating when] I had been single for too long. I got a feeling of lacking someone in my life and I wanted the feeling back where you feel that you are on top of the world! All of a sudden I decided to join this internet dating site [. . .]. I am looking for a serious relationship. I want a girl with whom I can advance hand-in-hand and whom I can trust; I know trust is the key to a sound steady relationship.
>
> (Male, 22 years old)

As the quote indicates, the future emotional state is clustered around the phantasmagoria of a heterosexual and monogamous relationship based on the idea of romantic love. Men talk about finding a woman and women talk about finding a man. They want a "steady relationship" with their future partner. This is called "being serious" on the Internet dating site, and is claimed and acknowledged to be more legitimate than looking for approval or making friends. Happiness thus comes to be associated with a specific form of phantasmagorical relationship. The phantasmagoria is also romanticized, especially among female participants, by using a code word for the future partner, such as "Mr. Right" or "the right one," as indicated in the following quote.

> I look for "the right one," but it is always nice to meet new people and being flirtatious. [. . .] Internet dating is about flirting and looking for love :).
>
> (Female, 27 years old)

Creating Phantasmagorias: The Front Page of the Internet Dating Site

> Excitement, buzzes, butterflies in your belly, desire, happiness, yes-no-maybe, hope. . . . This is where it starts. Hopeful meetings in a world full of emotions. There are thousands and thousands of singles here online—eager to find acquaintances, discuss whatever is important or unimportant to them and find a date. If you take the opportunity, there's a good chance that the adventure begins right here, right now.
>
> (Retrieved from the Internet dating site, April 12, 2010)

This quote is taken from the front page of the Internet dating site. It is part of the production and reproduction of feelings on the Internet dating site. The conditional basis for action is set by offering action possibilities through a narrative adjusted to the presumptive participants of the Internet dating site. The prescribed emotions in the narrative are not obligatory, but the narrative does acknowledge the most legitimate way of feeling at this site for action, i.e., its most legitimate emotional order. One is expected to act upon the narrative toward overcoming one's current life situation in light of this phantasmagoria. Hopeful meetings involve interactions with presumptive partners, such as making acquaintances, which can lead to a date. The uncertainty in Internet dating is taken into account. But hope keeps Internet dating going, because hope contains the possibility, although not the certainty, of finding someone.

Furthermore, on the front page of the Internet dating site a relatively large picture is displayed. The picture focuses on a woman. She is smiling while gazing intensely at a man looking down toward the picture's corner. The picture introduces the Internet dating site to the visitors as being about heterosexual relationships and happiness for both women and men. The picture evokes a possible and imaginable future. It creates an inbuilt emotional expectancy. It contributes to preparing the participants to feel hope and to interrelate toward achieving this end. If they do not already comply with the emotional ordering of this particular social world, they are led to comply with it. The picture can be interpreted as an "ordinary happy date." But it can also be about the state in which the participants enter when "the search is over," i.e., when they move into another arena together with whom they have "found." In both cases, it is about showing happiness in order to create hope in Internet dating.

The Role of the Internet Dating Companies in Shaping, Tuning, and Commercializing Internet Daters' Emotions

One of the roles of the Internet dating companies is to tune conditions for the arousing of specific emotions by means of the Internet dating site. *The Internet dating companies participate in the shaping, tuning, using, and evoking of the emotional ordering of the social world of Internet dating at this site for action.* It is a process of production and reproduction. I suggest that the role of the Internet dating company is to initially "fit" its reproduced version of the phantasmagoria of Internet dating to presumptive and existing Internet daters. If they are not in a state of hope, they could be made to feel hope by contrasting a future life of happiness with their current life situation. In other words, the company indirectly tunes the emotions of the persons on the Internet dating site by drawing upon phantasmagorias. The shared experience of hope is what the Internet dating companies make revenue from (see also Arvidsson 2006; Illouz 2007). People buy access to communication technologies for direct contact with other Internet daters. The Internet dating companies have an interest in keeping the Internet daters in a state of hope through stabilizing the emotional ordering, as hope is to be constantly present and to reproduce the (economically) valuable Internet dating activities. This endeavor is materialized in the social action that the Internet dating site enables and allows. Since the Internet dating site partakes in setting the conditions for action, it is hard to pursue other activities and have other ambitions. But the phantasmagorias of Internet dating are also reproduced by the interaction taking place through Internet dating on this Internet dating site.

The participants perform immaterial labor for the Internet dating company by imagining possible social relationships with others (Arvidsson 2006). Their activity constitutes emotional labor (Hochschild 1983). But this emotional labor does not result from company standards, as the participants are not employed by the company. They are prosumers (Ritzer and Jurgenson 2010): they consume what they produce together on the Internet dating site, mainly through their interaction with the site and with other members. Participants in Internet dating are seeking and getting approval and hope as part of this prosumption.

Emotional Socialization, Emotional Careers, and the Blasé Attitude

The following quote is an example of a person describing her emotional socialization and emotional career, i.e., her process of gaining control and ability to regulate her hope, and her commitment to Internet dating in search for a future romantic partner.

> I used to get that tingly feeling all the time in the beginning. I'd get my hopes up quickly, thinking I'd found my prince after just a few letters exchanged. But I learned pretty soon that it's often false hopes, and that you shouldn't tell the other person everything about yourself immediately, because suddenly he stops answering your letters. It doesn't happen all the time, but it does happen. So I'm afraid of hoping too much before getting to know the other person. Today I get belly butterflies, but not instantly. After writing quite a lot and really getting to know the person, the sensation that it could actually lead somewhere makes it happen.
>
> (Female, 20 years old)

As the participants come to be socialized into the world of Internet dating, they learn from experiences of Internet dating to regulate their hope. For example, they accept events as they unfold, become relaxed, and stay focused on the present rather than the future. This is the process of emotional socialization, which leads to the adaptation and regulation of expectations. The participants come to diverge from the most legitimate emotional order of hope, and experience both "too much hope" and "too little hope." The commitment to the Internet dating world is based on hope. Thus, being "too committed" to the Internet dating world is the same as having "too much hope," whereas having "too little hope" means being "less committed." The participants need to feel the right amount of hope in order to be able to conform to the most legitimate emotional order of the Internet dating world, as set by articulated phantasmagorias in the situation of action. The consequence of drifting into too much or too little hope is to become loosely bound to the Internet dating world and adopt a "relaxed" or perhaps, in Simmel's (1990) terms, a blasé attitude. This attitude can be paraphrased as being unconcerned with one's values and handling value differences through rationalization and formalization of human life, of which Internet dating is an example. Internet dating can be understood in these terms, as the activities on the Internet dating site are about weighting, calculating, numerical determination, precision, punctuality, and stability (Simmel 1990). In a context of overstimulation, the blasé attitude is an embodied attitude of indifference to what is going on (Simmel 1990: 256–257). The participants have thus suppressed their feelings of hope, which leads to this emotional attitude.

Realizing the Fragility of Social Relationships in Internet Dating

Participants looking for a future romantic partner strive for a life situation in accordance with the ideas of monogamous heterosexuality and romantic love. An emotional commitment of hope is invested in the other as being the means of realizing a phantasmagorical future. When the hopes are not met, the participants may feel disappointed and lose hope. People tend to come and go. The social relationships of the Internet dating world thus have a temporary character and hold people together until further notice (Bauman 2003).

> Internet dating can also be very trying and take a lot of time. I only answer letters from people that I feel a genuine solid interest in. It takes too much time to write letters of nonsense to people that don't even fit my profile. Also, it's very frustrating when "talking" to a guy for a while, building up some sort of friendship where both have shared things about life and personal stuff, etc., then actually meeting him and all of a sudden he's just not interesting anymore. And the other way round. All of a sudden this "friend" has vanished. It's no doubt hard to open up time and time again, sharing your qualities. Eventually you talk less.
>
> (Female, 40 years old)

This quote, similar to the previous one, illustrates the coming and going of people and the fragility of social relationships on the Internet dating site, due to the fact that Internet dating is based on the idea of evaluating the other as a potential partner. Eventually, if the participants do not get the approval they look for, this might result in the blasé attitude.

Generally, participants with a relaxed attitude are loosely committed to Internet dating, because Internet dating is about the future, and this future is hopeful but contingent. It is encouraging but also risky. It is a future to which one should hold on lightly in order not to be disappointed. Those participants are in a position between hope and having lost hope. They engage in a personal negotiation of the commitment to Internet dating. This emotional career of emotional socialization seems to be common to many participants who do not achieve their initial hope of finding "someone."

DISCUSSION

Searching for a future romantic partner generates a process of evaluation in Internet dating. This process is enhanced by the future-oriented and action-oriented emotion of hope for happiness, which can be tuned by phantasmagorias of a romantic monogamous and heterosexual relationship. The future is represented by talk about sought-for emotions. Practicing Internet dating involves an emotional socialization process, where people come to regulate their emotions of hope by reflective control-based ongoing experiences, such as realizing the fragility of social relationships. Participants can gain approval when they are evaluated as desirable on the Internet dating site. This approval can serve as a basis of hope that one might meet a future partner. But it can also be an end in itself.

The (re)production of Internet dating activities goes on as long as people are committed to the activities of Internet dating. This is primarily achieved by maintaining a state of hope. The (re)production of activities on the Internet dating site is based on this essential social process. The Internet dating companies play an active role in tuning people into a state of hope, since it makes them committed to Internet dating and turns them into producers of valuable content for the Internet dating site. The hope for happiness is the energizing social process that keeps Internet dating going. Happiness is a construct traversing discourses of heterosexuality, monogamy, and ideas of an idealized romantic love. Thus, the discursive framing puts pressure on individuals to achieve self-realization through these attributes. Happiness in these discursive forms is constitutive of people's actions in the present. Ideas about happiness thus play a crucial role in generating actions of self-realization and beliefs around life conditions. A critical self-reflection on the prevalent discourses of happiness might produce ideas of other options for self-realization.

The search for a future partner seems to be about creating certainty in the future, while the search itself and the participants' situation are often seen as uncertain and ambivalent. The search is oriented toward the social relationship with a future partner, and many seem to expect this relationship to signify happiness, certainty, and new plans for the future. One may thus see Internet dating as a space in between, a transitional sphere, where the intended transformation of how one sees oneself seems to be possible. The social activities involved in practicing Internet dating are intentional, temporary, and transformative. They are intentional since there is a goal: the search for a future partner. They are temporary and transformative since they are constructed as a way to overcome one's current life situation. Furthermore, as these activities appear to be intentional, temporary, and transformative, the ambition for each individual is always to make Internet dating activities unnecessary and to ultimately take part in some other activities. The social world of Internet dating, being based on the search for a future romantic partner, in fact strives for its own destruction. This particular collective action is destructive as the participants leave Internet dating when they have found what it is all about; the social world of Internet dating exists because people need it as a stepping stone. However, the Internet dating companies have an obvious interest in the continuation of Internet dating activities and in recruiting new members for their own reproduction. The Internet dating as such is upheld by the continuous activities of Internet dating, as well as by the technologies and discourses of the Internet dating site.

Further studies need to address emotional socialization and emotional orderings of social worlds related to the role of commercialization of transitional spaces, of which Internet dating is an example. Looking for a job might be yet another example. Further studies could also focus on a comparative analysis of Internet dating practices at different Internet dating sites, which so far has not been done.

ACKNOWLEDGMENTS

I would like to thank Patrik Aspers, Agnieszka Bron, Andreas Henriksson, and Franz Kernic for their comments to earlier drafts of this article. The chapter is based on the author's master's thesis. An earlier version was published as Fürst, H. (2011) Overcharged Emotion: Internet Dating, Negotiation and Emotion. *Terazniejszosc Czlowiek Edukacja*, 56(4), 125–137. Reprinted by permission.

NOTES

1. A Swedish survey study conducted in 2009 reveals that most respondents who had met a partner during the last four years had initiated their latest partner relationship on an Internet dating site (Forskning and Framsteg, n.d.).

REFERENCES

Arvidsson, A. (2006) "Quality Singles": Internet Dating and the Work of Fantasy. *New Media & Society*, 8(4), 671–690.
Bauman, Z. (2003) *Liquid Love: On the Frailty of Human Bonds*. Cambridge, UK: Polity Press.
Becker, H.S. (1982) *Art Worlds*. Berkeley, CA: University of California Press.
Blumer, H. (1969) *Symbolic Interactionism: Perspective and Method*. Englewood Cliffs, NJ: Prentice-Hall.
Clarke, A. (1991) Social Worlds/Arenas Theory as Organizational Theory. In A.L. Strauss and D.R. Maines (Eds.), *Social Organization and Social Process: Essays in Honor of Anselm Strauss* (pp. 119–158). New York: A. de Gruyter.
Clarke, A. (2005) *Situational Analysis: Grounded Theory After the Postmodern Turn*. Thousand Oaks, CA: Sage Publications.
Couch, D. and Liamputtong, P. (2008) Online Dating and Mating: The Use of the Internet to Meet Sexual Partners. *Qualitative Health Research*, 18(2), 268–279.
Daneback, K., Månsson, S. and Ross, M. (2007) Using the Internet to Find Offline Sex Partners. *CyberPsychology & Behavior*, 10(1), 100–107.
Denzin, N. (1983) A Note on Emotionality, Self, and Interaction. *American Journal of Sociology*, 89(2), 402–409.
Denzin, N. (1984) Reply to Baldwin. *American Journal of Sociology*, 90(2), 422–427.
Engdahl, E. (2005) *A Theory of the Emotional Self: From the Standpoint of a Neo-Meadian*. Örebro: Örebro University.

Forskning and Framsteg (n.d.) *Störst av allt är nätet* [Biggest of All is the Internet]. Retrieved April 26, 2011 from http://www.fof.se/textruta/storst-av-allt-ar-natet

Glaser, B.G. (1978) *Theoretical Sensitivity: Advances in the Methodology of Grounded Theory*. Mill Valley, CA: Sociology Press.

Hancock, J.T., Toma, C. and Ellison, N. (2007) The Truth About Lying in Online Dating Profiles. *Proceedings of the SIGCHI Conference on Human Factors in Computing Systems*, 449–452.

Hardey, M. (2004) Mediated Relationships. *Information, Communication & Society*, 7(2), 207–222.

Hochschild, A.R. (1983) *The Managed Heart: Commercialization of Human Feeling*. Berkeley, CA: University of California Press.

Illouz, E. (2007) *Cold Intimacies: The Making of Emotional Capitalism*. Cambridge, UK: Polity Press.

Lawson, H.M. and Leck, K. (2006) Dynamics of Internet Dating. *Social Science Computer Review*, 24(2), 189–208.

Luckmann, B. (1978) The Small Life-Worlds of Modern Man. In T. Luckmann (Ed.), *Phenomenology and Sociology* (pp. 275–290). Harmondsworth, UK: Penguin.

Mead, G.H. (1910) What Social Objects Must Psychology Presuppose? *Journal of Philosophy, Psychology and Scientific Methods*, 7, 174–180.

Mead, G.H. (1982) *The Individual and the Social Self: Unpublished Essays by G. H. Mead* (D. L. Miller, Ed.). Chicago, IL: University of Chicago Press.

Mead, G.H. (2001) *Essays in Social Psychology*. New Brunswick, NJ: Transaction Publishers.

Ritzer, G. and Jurgenson, N. (2010) Production, Consumption, Prosumption: The Nature of Capitalism in the Age of the Digital "Prosumer". Journal of Consumer Culture, 10(1), 13–36.

Simmel, G. (1990) *The Philosophy of Money*. London: Routledge.

Strauss, A.L. (1978) A Social World Perspective. *Studies in Symbolic Interaction*, 1, 119–128.

Strauss, A.L. (1982) Social Worlds and Legitimization Processes. *Studies in Symbolic Interaction*, 4, 171–190.

Strauss, A.L. (1984) Social Worlds and Their Segmentation Processes. *Studies in Symbolic Interaction*, 5, 123–139.

Strauss, A.L. (1993) *Continual Permutations of Action*. New Brunswick, NJ: AldineTransaction.

Whitty, M.T. (2007) Revealing the "Real" Me, Searching For the "Actual" You: Presentations of Self on an Internet Dating Site. *Computers in Human Behavior*, 24(4), 1707–1723.

von Wright, M. (2000) *Vad eller vem?: en pedagogisk rekonstruktion av G. H. Meads teori om människors intersubjektivitet*. [What or Who? A Pedagogical Reconstruction of G. H. Mead's Theory on Human Intersubjectivity]. Göteborg: Daidalos.

7 Emotion to Action?
Deconstructing the Ontological Politics of the "Like" Button

Tamara Peyton

> She totally liked him on Facebook so that she could track what we were saying. But, like . . . she doesn't really like him, ya know? Liking him on Facebook doesn't mean she 'like likes' him.
>
> (overheard on a city bus)

THE LIKE BUTTON: AN INTRODUCTION

The 'like' feature was added to Facebook in February 2009. In just under four years, it has become one of the most ubiquitous sociotechnical objects on the web. Practically replacing the former 'share' idea, the like button has eclipsed most other methods of sharing content and has become a metaphor for information sharing that has spread outside of Facebook to other social media. Statistics from the Facebook corporation itself (2010) states that content clickthroughs on 'liked' content appearing on Facebook has increased traffic to major web content hubs by 150% to 500%. The Facebook corporation suggests that via Facebook or elsewhere on the web, the like button is threatening to overtake tweets as the predominant metaphor and method for the social sharing of digital media. But what is this like button? What does it mean to like something or someone? What is going on here?

Inspired by that conversation I overheard on the bus, what I am arguing in this chapter is that the commonly understood idea of 'liking' has undergone a semiotic shift in both meaning and intent as a result of the Facebook like button. What it means to 'like' has moved away from the realm of the emotive internal life of individuals and into the realm of the discursive public sphere of societies. Instead of being tied to an internal sensation that reacts tacitly to an external stimulus, to 'like' now becomes a conscious rationalized action that connotes an external tag of connection between an individual, a discursive element, and a social stance. Whether the act of clicking 'like' is done to a Facebook status update, to a website forum post, or to a YouTube video, to 'like' is now to act, rather than to feel.

My objective in this chapter is to problematize the idea of digital 'liking,' by tracing out the action orientations of the like button via four case studies. I am questioning whether or not 'liking' digital content is still congruent with the idea of the commonly understood expression of a personal emotion. The point is to have us think critically about what we think liking is, what we believe we are doing when we demonstrate digital liking, and what the digital liking is doing for us and to us as social actors and neoliberal consumers.

UNDERSTANDING "LIKING" THEN AND NOW

According to the OED, the classic definition of the term 'like' means "to take pleasure" (OED 2011: 2). Noting that the word is undergoing a transition of meaning in current times, the OED says also that to like is "to find agreeable or congenial; to feel attracted to or favourably impressed by (a person); to have a taste or fancy for, take pleasure in (a thing, an action, a condition, etc.)." (OED 2011: 6). This gets closer to the more common understanding of liking as an emotion. It is instructive, but not entirely helpful. As the OED notes, this transition in meaning has the concept behind the word undergoing a semiotic shift, a shift that is still underway, making it difficult to attempt to pin it down and fix it into a crystallized meaning.

By then delving into the sociology of emotions literature, I realized that the original social idea of 'liking' was not really of liking as emotion either. If, as Thoits (1989: 318) argues, emotions are "culturally delineated types of feelings or affects," this does not quite match the action orientation of Facebook's digital 'like.' Instead, in its tacit and explicit meanings, Facebook 'liking' is more aligned to Gordon's idea of a sentiment; a "socially constructed pattern of sensations, expressive gestures, and cultural meanings organized around a relationship to a social object" (Gordon 1981: 566). But the original and ideal meaning of liking as sentiment is not particularly helpful, because insomuch as the idea of 'liking' something or someone is commonly understood to be a feeling, and feelings are considered akin to emotions in popular understanding, thus, liking is nevertheless considered to be feeling, an interpretation shored up by the OED's notation of the term's in-progress meaning shift.

DIFFÉRANCE

What this highlights is the idea that translation is always at work in meaning making. In order to understand the variability and transitional nature of language in action, I have to attend to politics of translation. Consequently, what is at work here is a great example of Derrida's (1982, 1998) thorny term *différance*, and the politics of deconstruction. Drawing from Saussure's idea that "in language there are only differences without positive terms" (Saussure 1959: 120, original emphasis), Derrida positions *différance* as an unstable patterning of transitory knowledge which is uneasily coalesced into an amorphous theoretical understanding that is very difficult to represent in speech,

Emotion to Action? 115

writing, or imagery. The root term is the contradictory French verb *"differer."* The French root term means both "to differ or vary" and "to defer." Derrida takes this contradictory sense of both the deferral of meaning and the differing meanings of things and ideas and embeds it into *différance*. But he cautions us, *"différance* is literally neither a word nor a concept" (Derrida 1982: 3) that has no actual or inherent meaning. Instead, it is a "neographism" (Derrida 1982: 3), or a condition or situation of possible meanings. He chooses to consider *différance* as a neographism rather than a neologism, because for him, the entire purpose of looking at a politics of translation via *différance* is to render unstable any linguistic or literary word, concept, or logic. Calling it a neographism shows the way in which the meaning of a visual thing (a symbol, a word, a graph, an image) can be practically impossible to put into words or capture in writing in any meaningful, stable way. The disruption and breakage of any sort of formal logic of intelligibility, and the dehistoricization of this interrupted sense of meaning, is a key political point for a deconstructive project. In the destabilization of the meaning of a word, concept, or thing, the politics of *différance* enables the possibility of making a multiplicity visible and of highlighting the flows of power circulating in everyday discourse.

POLITICS OF INTELLIGIBILITY OF THE THUMBS-UP BUTTON

Thinking through the politics of *différance*, the like button as a sociotechnical feature of digital life carries a variety of politicized meanings and understandings. A kind of meaning register shift occurs in the digital lifeworld via the interactions initiated and extended through the use of the thumbs up icon with its associated word 'like.' The logic of intelligibility is disrupted through the exteriorization of previously interior perceptions and ideas. Liking moves here from a common parlance sense of 'like' as a tacit amorphous emotion or sentiment to an active and externalized expressive action made explicit via the like button. This shift in register is equally a shift in temporal status. Time then becomes one of the political agents of meaning making around the thumbs up button. To be known as a 'liker' requires a waiting period that occurs between the act of clicking the 'like' button and the reading by others of that 'liking' action. In that liminal period between clicking the like button and being perceived as a liker, the button exteriorizes the necessity of the anxiety of waiting to the liking user. "Will people be happy that I like this thing? How will my friends react?" are the kinds of things that happen internally in the liker while they wait for the reaction to their reaction. In that waiting process, there is a constant sensation of instability. The liking user experiences a sense of things not being understood yet or resolved. The resolution to that liminality occurs when a liker's social circle notices their like and is thereby coerced into reacting. Even if the second person makes the choice to not do anything (i.e., to not 'like' the item, too), that conscious choice is a kind of tacit knowledge mobilized into explicit and situated action. More frequently, a chain of liking occurs within one's social circle, connecting individuals and reinforcing the demonstrative social action power of the like button.

As the idea of liking circulates out from the individual's social circle, it moves into the economic-primary world of the commercial, shifting via a black boxing strategy. In digital technology, Galloway (2001) notes that the idea of a 'black box' is that of a complex technological system made to appear to the end user as a simple, easy-to-use object. A variety of strategies are employed to mask or hide the complexity of the system in order to keep users feeling a sense of mastery over their technological tool. Thus, black boxing is a strategy and technique simultaneously that is wrapped up in the complexities of neoliberal capitalism. One of the most prominent and successful promoters of technological black boxing is Apple Inc. The clean lines and deceptively simple nature of their computer systems and portable devices allow users to feel in control of their technical toys, and Apple's products require little technical knowledge to operate. Both in its materiality and in its intellectual force within the technical lifeworld, the entire complexity of an Apple computer system is made to fade into the background.

The same is true of the like button. The complex flows of networked code, financial logic, consumerist aspirations, and informational media get wrapped up and blackboxed in the small, simplistic, and seemingly universal symbol of the thumbs-up button. The like button is a chimera, meaning multiple things to many people, and acting within the sociotechnical lifeworld in a variety of ways based on the socio-political position of the button publisher, the invoking user, and the neoliberal economic practices of information consumption in which it is embedded.

Given that social media sites are ostensibly free for users, the like button figures heavily in these social media sites' market capitalization. To these companies, the idea of liking is oriented toward the advertising and investment market. The quantity of those who have liked an item is used by these companies as an indicator of loyal and receptive fans for another company's products. As packaged into a consumerist discourse, it is sold by the social media platforms to other companies and political entities. Via this register shift, to like is no longer just a personal emotion or a social expression. Instead, it becomes finalized as an active capitalist stance, a demonstration of consumer alignment toward specific market commodities. The action of 'liking' in the digital context is part of a flow of ontological politics bound up in neoliberal market capitalism.

THE TRANSITIONS AND SHIFTS

This liking has three component shifts or phases in it. Moving from an individual reaction to a social display of association and then into an indicator of socioeconomic affiliation that can be leveraged for financial and political gain, liking now underpins the entire economy of social media and digital networks. To like something has ceased to be an emotion; that much is clear. But it is impossible to come up with a singular meaning now of what liking is. It has multiple meanings, multiple possibilities. It has become instead (variably) a tracking device, a browser bookmark replacement, a vote, a demonstrator of solidarity, a peer pressure indicator, a contest entry, a content aggregator, a kind of social currency . . . the list goes on. In the strong binarisms that infuse marketing speak, the like button on digital content is considered to be a call to action. It is an encouragement, an incentive, a blandishment. When each of these types of 'liking' are examined, in all of these manifestations and interpretations, it becomes apparent that there is little to be said about liking as a meta concept. As simultaneously expressed and nullified in the word 'like' with its thumbs-up button, 'liking' is a black box of semiotic and semantic possibility. I can grasp at potential possibilities presented via digital liking, but I cannot (and should not) attempt to fix any one of them. Doing so would be a political act of sabotage on the possibilities inherent to the concept as it now circulates in digitality. In our neoliberal market-dominated world, liking exists as part of digitized *différance*.

CASE STUDIES: FROM "SCIENCE" TO RETAIL TO GAMING TO WEBSITES

I spend a lot of time online each week, navigating websites, looking up information, keeping in touch with the interests of family, friends, and colleagues via Facebook, Twitter, and Pinterest, and looking for content to use for teaching. Consequently, over the period of two months that I spent documenting every instance I noticed of the like button, I captured over 150 images. This quantity does not include every time it appeared on every Facebook page, info stream, or comment cycle, nor does it account for the many times it appeared on forum posts on the website of one of my current favorite video games, or on the various YouTube channels I track each week. Also, I discovered that once I started getting attentive to the little blue and white thumbs-up picture, I found it literally everywhere[1]. This small thing had been part of my digital world for the last handful of years, yet I rarely bothered to notice its presence. Only once I set out to explicitly find and capture it did I realize just how ubiquitous it is, and how odd and varied its actual meaning was, when considered in flows of situated action.

It is impossible to present here every single case of the like button that I screen-captured. I have to be selective. To illustrate this variability of action around digital liking, I have chosen to highlight four different types of liking chains of action via anonymized[2] descriptive case studies.

The first case study is the sharing of a science news story. The second is an invitation to like a commercial product on a website. The third is the suggestion to like a YouTube video channel about an online video game, and the final case study is an invitation to like two different retail companies' Facebook pages. These selections are somewhat arbitrary, and I will stress they are just a few examples of the over 150 case studies I tracked and documented. I have selected them because of the way their particular employment of the idea of liking highlights some of the key politics of *différance* around digital liking as neoliberal action. Following Stake's (1994) case study typologies, the selection of these four cases act as a blended case study that represents both the study of the particular and of the general instrumental features of each case. When considered collectively, these cases highlight the politics of *différance* of liking in the digital context.

MAKING THE CASE #1: SCIENCE "NEWS"

An academic colleague of mine apparently found an online science-themed magazine article interesting and 'liked' it to his Facebook profile. Within an hour, a half dozen of us had read the liked article and commented on it back on his Facebook post. Another two people clicked the 'like' feature below the Facebook master post, with a few other different people clicking that they 'liked' a comment made by someone in the comment flow.

While at first this appears to be a fairly normal and mundane act that happens all over the net, thousands of times daily, what makes this particular case interesting is both the content of that article and the date on which it was posted. The title of the science-themed article was 'Masturbation calms restless leg syndrome' and the byline for the article lists its posting date as April 1, 2011. My colleague simply tagged this 'liked' piece as being 'great news!' The rest of us weighed in with a variety of comments and liking actions. But at no point did anyone make it clear whether or not they thought the article was real, and what they meant when they clicked 'like.' Were they saying they thought the article was true and if so, it was indeed great news? Were they saying that they considered the whole thing an April Fool's joke by the magazine, and so by clicking 'like' they were in on the joke? Did they click 'like' because they found it funny? Because they wish it were true? Because they like masturbation? Because they had no clue what to make of it nor did they know what else to do but they still wanted my colleague to know they had noticed it? If one of the downstream commenter's friends noticed the 'like' notation on the friend's profile about this article, and then 'liked' that friend's 'like,' what does this mean? Given that the additional liking in that last instance would happen only on that commenter's wall, the rest of the downstream would not be aware of it.

Via this example, I see the muddiness and uncertainty that the 'like' button brings to social media. The apparently explicit meaning of the button, to literally 'like' content in this case, gets blurred out and disappeared in the stream of possibilities of unfixed meaning, through the liking actions.

In this example, digital 'liking' exhibits a tension between authenticity and seriousness, and between ambiguity and humor. Being unable to determine if the content of a magazine article was actually authentic or not, yet sharing it anyway with no stance taken toward the truth or humor potential of it, my colleague who 'liked' it to his Facebook profile initiated an event that moved the possibility for its truth or fiction, seriousness or humor further away from the realm of possible knowledge. And we all followed along down that stream.

MAKING THE CASE #2: ELECTRONICS PRODUCT

My reasonably new 22" LCD monitor was getting increasingly foggy, muddied up from fingerprints and dust and the occasional splatter of coffee or tomato sauce from mornings and evenings spent toiling away in front of my computer. I needed to clean it, and it requires a special product to do so. Being a fan of not leaving the house unless I have to, especially on a Sunday, I fired up my browser and went to a major electronics retailer's website. There, I found Anti-Static LCD Monitor Wipes. At $12.99 for 80 wipes, they seemed a good buy, so I added them to my cart and went to complete my purchase. However, the system glitched and would not take my U.S. credit card. I would have to go to the store to purchase them.

Rather than bookmarking the page and trying to remember where I put it in my burgeoning pile of bookmarked content in my browser, I chose to click that little like button so that I would have it on my Facebook to remember later, just in case the closest store did not have them on hand. I got on the subway, went to the store, found and purchased them, and came home. I then got busy with other things and did not get around to using them for five days.

When I finally remembered my intention to bring my monitor back to its out-of-the-box state of cleanliness, I grabbed the tube, pulled out a wipe, and set to cleaning my monitor. At least, I attempted to do so. The wipes did not work well. While they removed the coffee and tomato splashes, they left little bits of whitish lint everywhere and got embedded in the corners between the screen and the frame.

What is the point of relating these details? There I was, with a faulty product that did not work well, but which I could not return. Yet I had clicked that I 'liked' them to my Facebook profile in order to record the product on my Facebook as a memory aid. I spent 15 minutes scrolling through almost a week's worth of Facebook activity on my profile until I finally found that 'liking' notation, and then I deleted it. One problem: I have no idea if doing so removes it from the retailer's website. Does the retailer still think I like these wipes? How can I make the connection for them between what I did when I clicked the button, what I really meant when I clicked like, and the fact that I now do not like this product and have rescinded my 'liking' action? How do I know if my 'liking' them originally now means that the retailer will think these are a good product that meets my needs and will use this 'liking' as a way of suggesting other products to me that align to this? And what if they sell my 'liking' action back to the wipes manufacturer's company? In my opinion, the product needs to be reformulated, because it is counter-productive to its intended purpose. However, by clicking 'like' on its retailer's product page, I have explicitly indicated to the makers of the wipes, via their retailer, that their product is just great as it is. This is now untrue for me, yet I have no way of adjusting that perception. My 'like' is not true anymore for me (and it never did mean that I emotively 'like-liked' the wipes in the first place) but it remains forever true to both the manufacturer and their major retailer.

MAKING THE CASE #3: LEAGUE OF LEGENDS CONTENT

League of Legends (LoL) is a free-to-play multiplayer online battle arena game (MOBA) in which players (known as 'Summoners') choose a champion to control, then team up in groups of five to compete against another team in a single 'capture the flag'-style match. The makers and promoters of LoL, Riot Games, are very good at keeping the LoL gaming community interested and invested in the game. They do this by promoting the paid in-game items and out-of-game merchandise that are their cash flow, by promoting the activities of players around the game, and by introducing a steady flow of new characters (approximately two each month) that must be purchased with actual currency.

Riot understands that the online community is where the interest is formed, supported, and extended. Consequently, they invest a lot of time each week putting various Riot employees front and center to their gamer audience. As part of their YouTube strategy, Riot publishes weekly 'Summoner Showcases.' In each of these, the audience is exposed to a wide variety of player/fan created content around the LoL game and gameworld, be it comics, rap songs, dance routines, cartoons, mock fights, or costume-play. Each Summoner Showcase is hosted by LoL official YouTube personality, Nikasaur. At the end of some of their Summoner Showcase videos, Nikasaur exhorts all viewers to "Subscribe to our YouTube channel via the feisty little thumbs-up button." Accompanying this entreaty with vigorous thumbs-up seesaw motions, Nikasaur wants us to like the video as a way of bookmarking this official LoL venue. But is that what the videostream's audience is actually doing?

Comments below many of these LoL videos suggest that is not the point to clicking 'like' in this situation. A large number of the people who 'like' the LoL Summoner Showcases do it to comment on the sexual appeal of Nikasaur as a female and as a gamer. A correlating reason for 'liking' actions around these showcases are to express solidarity or a sense of appreciation

Emotion to Action? 121

Figure 7.1 Riot Games personality Nikasaur encourages viewers to 'like' their video feed.

for the debased shared humor that accompanies the outrageous sexual comments made by some players. One other reason demonstrated for 'liking' these videos is to highlight some portion of the Nikasaur's onscreen actions and to draw attention to them as gendered action. Very few of the comments around 'liking' action seem to be to indicate that they liked the content of the videos, or that they liked any of the changes that Nikasaur may have highlighted or discussed.

As shown in another LoL YouTube case, the 'liking' is less about the content of the video itself as presented by Riot, and more about the performance of a kind of player identity that is wrapped up in male humor, posturing, demonstration, and attention-getting. It is a community of gendered practice around gaming. As McConnell-Ginet describes such identity demonstrations, "social identities, including gendered identities, arise primarily from articulating memberships in different communities of practice" (McConnell-Ginet 2005: 71).

In this example, the video was the 'teaser' or trailer to the game, and was released prior to its launch. Game release trailers are similar to movie

Figure 7.2 Official promotional video for Riot Games' League of Legends.

release trailers in that they are created and distributed with the intent of giving players an idea of what the gameworld will look and feel like, what the game play will be about and what the point to the game is.

In the kind of reverse temporality that happens around 'liking' actions, what has occurred around this LoL trailer video is 'likings' and accompanying comments from players who have already played the game and who draw on an extensive game knowledge. 'Liking' here seems to not be about the eagerness and suspense of waiting for the game to come out and finding out if it is fun to play. Nor is it based on speculative comments about what the game will be about, what happened in the betas, etc. Instead, the comments here are related to mundane points of lore from the video that the players feel have been misrepresented by Riot. Some 'likes' and comments deal with the strong difference in visuals between what is shown in the videos and what you actually get when you play the game. Others deal with the fact that Riot dropped the add-on name of 'Clash of Fates' from the official release title of the game. Overall, the comments appear to work more to demonstrate a sort of herd mentality of shared male gamers and their 'in the know, now!' state of play.

MAKING THE CASE #4: HOME GOODS RETAILER

I have demonstrated thus far that 'liking' is bound up in personal and social politics of *différance*. The case studies I have already presented highlight and problematize the idea of 'liking' as an authentic personal expression, and/or as a social alignment within a specific group. But there is one other kind of politics of 'liking' I would like to discuss and critique; that of economic 'liking.' In capitalist 'liking' chains, the 'liking' action is bound up in the consumerist politics of our neoliberal context. While neoliberalism has many varied meanings and interpretations,[3] the theme that runs through most work on neoliberalism is the idea that it is a highly rationalized political force and social ethos that shapes actions and conducts in the modern realm (Paradis-Peyton 2010). As a central preoccupation, Harvey (2005) notes that the freedom is often touted as a central neoliberal tenet, but it is a twisted and stunted freedom that ties itself to an economic predetermination and to the notion that the market will be the focus and source of an individual's self-actualization. Thus, economic control is the ultimate source of political power in neoliberal systems of culture. Neoliberal culture frames notions of human ability and individuality via expressly capitalist economic action and potential.

In this economic context, Facebook 'liking' becomes a consumer recruitment and management campaign. To demonstrate this, allow me tell you a story of an online retail experience I had with a major U.S.-based home goods retailer. Emphasizing clean lines, neutral colors, middle-class quality and comfort, strong design tendencies, and savvy home arranging, the retailer is one of the top companies in its economic strata.

Surfing over to the retailer's website one day, I was greeted not just by its familiar neutral colored palette. Within two seconds of arrival, a little box popped up, graying out everything below it. It urged me to "like us on Facebook and enter for a chance to win $10,000." Just in case the dollar amount was insufficient, if I did choose to 'like' them, I would also get the chance to have one of their stylists in my home "to offer advice on your makeover."

Note the phrasing of the invitation here. The "like us on Facebook" language highlights the action invocation of the idea of 'liking.' While intended to be active sounding and inviting, the declarative commanding tone is also placed first, so that it is not watered down by the idea that you only have a "chance to win" the $10,000. It was also interesting to see that nowhere does the little blue thumbs-up icon appear in this inviting box. Is this a suggestion that it would be too gauche to do so for the retailer's clientele? Using the thumbs-up would be the wrong cultural code for the retailer's upper-middle-class clientele perhaps? It suggests to me that excluding that button sets them apart from their competitors. So is this simply another example of a common retail desire to appear unique and different? Granted, the thumbs-up button appeared at the bottom right corner of every page of the retailer's site, but it was small and quite discreet, though positioned directly next to an invitation to "Be Social!," and also displayed a tally of how many people had clicked the thumbs-up button. This appearance on every page is odd, because the enormous number of 'likers' attributed to it is not tied to specific products or website features. The number of 'likers' is the same on every page. What do they signify?

Considering this bland ubiquity against the offering to win money and a makeover if I click "Enter Now" and allow my 'like' of the retailer to appear on my Facebook profile page, I see a strange duality of intent around digital liking. On the one thumb, the act of Facebook liking is something that is important enough to this home goods retailer for it to offer a large sum of money and a prestigious free service in return for having me do so. On the other thumb, Facebook liking overall on the home goods website is generic, decontextualized, and practically meaningless.

So what is Facebook liking to this major retailer? Based on this example, in its overt sense, it is simply a marketing tag, an indicator of a consumer's alliance with the company and its brand image. But the tacit part of this is the fact that once liked, the home goods retailer gains the right via the Facebook technological protocol and platform to track certain actions, and to use the things that I do in and around Facebook to tailor its offerings and to expand and deepen its explicit understandings of my everyday digital movements and doings.

ONTOLOGICAL POLITICS OF LIKING

The four case studies I have presented demonstrate that the thumbs-up button and the liking action are shifting toward an orientation aligned predominately with Facebook. This suggests that a variety of power plays and social relations are at work. To consider this question of social power and its contestations, I take guidance from Mol's (1999) work on *ontological politics* in actor networks. Ontological politics are the package of acts and actions within an environment that strive to shape the realities of action within the environment. Ontological politics are therefore wrapped around the neoliberal mobilization of knowledges that digital spaces such as Facebook make possible and control. The actions and uses of liking in Facebook meta-space demonstrate the ways in which tacit knowledge, once mobilized into explicit action, work against the formal and idealistic understandings of plans of action spoken of by Suchman (1987). Instead of individualized emotion, liking becomes neoliberal situated action. As a strategy of political consumerist power, the neoliberal action of Facebook liking maintains vestiges of the tacit but atrophied ideas of liking as emotional sentiment, while wrapping it up in the newly explicit sense of liking as action and stance toward capitalist products.

DIGITAL LIKING AS KNOWLEDGE GENRES

Influenced by Mol, analyzing these digital 'liking' practices via tracing the path of liking is intended to consider "how a specific version of the truth got crafted, what supported it, what was against it, and how its likely alternatives got discredited" (Mol 1999: 76). Through my case study data, my central concern has been to ponder the ways in which 'liking' is undergoing a meaning shift. Across various sites and within various social circles, how does the like button perform as an action object? How does it externalize and reshape the tacit idea of liking as an emotion into an explicit and multilayered situated act, embedded in specific neoliberal capitalist contexts? The answer to this is found in the like button's capacity to transform locally specific situated knowledge into globally capitalist Facebook social action.

Mol argues that action is implicitly and explicitly performative. This bears out in the case studies I have presented, and in the dozens of other interactions with like buttons I have had, during this study and beyond. Focusing on the performativity of liking necessitates an attention to the different versions of the liking action object being invoked simultaneously by the knowledge of and mobilization of the liking act via the button. It requires paying attention to the multiplicity of the interpretative meaning of a single action object as it plays out in action within a context.

In this sense, then, following Mol's ideas of knowledge genres, understanding the multiple performativities of knowledge which help shape knowledge genres has nudged me toward attentiveness to the multiplicities of possible tacit and explicit knowledges circulating around the like button and embedded in its appearances and uses. While Mol's ideas of knowledge genres are grounded in medical knowledge genres, they can be easily and faithfully transposed to other realms. I use her idea of knowledge multiplicities to consider the ways in which the circulation of liking acts upon the individual using the button, the audience who perceives that act, and the digital commercial environment in which the action takes place.

If we understand digital liking to be bound up in action, then we can understand the tacit and explicit meanings of digital liking to not be solely about its plan, in the sense of planning that Suchman describes. Its plan is the formal idea of a term. Yet the exteriorized formality of liking-as-emotion does not necessarily inhere in its individual usage or in its social and capitalist interpretation. As such, the reality of a liking understanding and a liking act is inherently multiple, and is thus implicated in multiple knowledge genres via situated action.

I employ Mol's genre concept to indicate a category of meaning understood either tacitly or explicitly in one of the realms of performative actions. Using these as a way to look at the case study data I have presented herein and across the other instances of digital liking I catalogued, I find some similarities and a few differences. I have discussed these as the political demonstrations of the personal, social, and economic spheres of neoliberal life. Within each of these spheres, there are meanings that I find correspond rather well to Mol's knowledge genres. Taking her genres and transposing them to my topic, I note Mol's formal genre is roughly equivalent to the conventional understanding of 'to like' something or someone as being an emotive act. Her intentional genre maps fairly well to the intended political purpose of use of any liking button. In this genre, liking is wrapped around the action idea of a demonstration of liking. Via a 'like' button, I am encouraged to take my internal and unexpressed emotion and make it *become* in the world, to cause my personal liking to be visible and to circulate externally and explicitly in the social realm. But, given the genesis of this action circulation, bound up in a commercial digital environment, there is inevitably also a situated genre. In the register of this genre, there occurs a mobilized interpretation of the liking action that is grounded in the variable social, political, and economic realities of action surrounding and threading through the liking chain.

As the case studies demonstrate, there is little sense left of any idea of Facebook liking as being tied to the formal genre of performative emotive action. Facebook or digital liking more generally is not the traditional idea of emotive liking. This is not an emotional response to a digital stimulus. Instead, there has been a register shift in meaning around digital liking. Digital liking is bound up in a tension between the intentional genre and the formal genre. Which register appears more dominant depends on a number of factors, including who initiates the liking chain, what the tacit purpose was in doing so, and the ultimate outcome of participating in liking actions. It is in this way that the three registers are threaded through with the individual, social, and economic ontological politics of digitality. The liking button is a demonstration of what Thoits calls "affective commitments" (Thoits 1989: 317). Digital liking is a way of reframing emotional identification and investment within a social network site into a new kind of affective commitment of consumerist and capitalist action orientations, embedded in a neoliberal context. When matched to an understanding of the fluid political nature of knowledge genres, the results of the collective case study method I have used highlight the ways in which digital knowledge production is rarely tacit. Flowing through the various realms of digital life, the possibilities of liking action are made multiple, problematic, contested, and contradictory.

Via this knowledge mobilization, this putting into play of Facebook action via the clicking of 'like' somewhere on the web, I have argued and demonstrated that the formerly understood private emotive capacity of liking has been reshaped into a sociotechnical actor. This has been borne out and deconstructed within the case presentations I have made. When the like button is included as an option on content, it becomes part of a networked assemblage of demonstrative capacity that can be read and then used in multiple ways. Remobilized into an explicit status indicator which functions both as a demonstrator of one's capacity to align preference toward content, or used to indicate solidarity with another's social stance, to 'like' is to make explicit one's liking. To 'like' is to act, rather than to feel. Digital 'liking' becomes the stimulation of a chain of actions and connections demonstrated digitally within the circulation of neoliberal discourses of socioeconomic action and preference.

NOTES

1. I do mean everywhere, outside of the obvious digital realm. I saw it on billboard advertisements on public transit. I found it in women's magazines, attached to both advertising and social issue awareness campaigns. I saw it represented as graffiti on a campus bathroom stall, and per the quote with which I started this chapter, I overheard it invoked in general conversation.
2. For copyright reasons, the images I gathered were not able to be included herein, and all descriptions have been anonymized.
3. Neoliberalism is considered variably to be: a theory of rationalized economic practices (Dean 1999; Harvey 2005; Mitchell 2002); an approach to the management of state, national, and private interests (Rose 1999); a pragmatist philosophy toward freedom and governance (Harvey 2005; Mitchell 2002; Rose 1999); a reaction to the left-tendencies of liberalism (Kendall 2003; Rose 1999); or a technology of power and identity management (Brown 2006; Ong 2006; Rose 1999).

REFERENCES

Brown, W. (2006) *Regulating Aversion: Tolerance in the Age of Identity and Empire*. Princeton, NJ: Princeton University Press.
Dean, M. (1999) *Governmentality: Power and Rule in Modern Society*. Thousand Oaks, CA: Sage.
Derrida, J. (1982) "Différance." *Margins of Philosophy*. Chicago, IL: University of Chicago Press.
Derrida, J. (1998) *Of Grammatology*. Baltimore, MD: Johns Hopkins University Press.
Facebook. (2010, September 29) *The Value of a liker* [online]. Retrieved April 1, 2011 from http://www.facebook.com/note.php?note_id=150630338305797
Galloway, A.R. (2001) Protocol, or How Control Exists After Decentralization. *Rethinking Marxism*, 13(3/4), 81–88.
Gordon, S.L. (1981) The Sociology of Sentiments and Emotions. In M. Rosenberg and R.H. Turner (Eds.), *Social Psychology: Sociological Perspectives* (pp. 562–592). New York: Basic Books.
Harvey, D. (2005) *A Brief History of Neoliberalism*. New York: Oxford University Press.
Kendall, G. (2003) *From Liberalism to Neoliberalism*. Paper presented to the Social Change in the 21st Century conference, Brisbane: Queensland University of Technology.
McConnell-Ginet, S. (2005) What's in a Name? Social Labelling and Gender Practices. In J. Holmes and M. Meyerhoff (Eds.), *The Handbook of Language and Gender*. Malden, MA: Blackwell.
Mitchell, T. (2002) *Rule of Experts: Egypt, Techno-Politics, Modernity*. Los Angeles: University of California Press.

Mol, A. (1999) Ontological Politics: A Word and Some Questions. In J. Law and J. Hassard (Eds.), *Actor Network Theory and After* (pp.74–89). Malden, MA: Blackwell.

Ong, A. (2006) *Neoliberalism as Exception. Mutations in Citizenship and Sovereignty.* Durham, NC: Duke University Press.

Oxford English Dictionary Online. (2011) *Electronic Resource* [online]. Retrieved April 1, 2011 from http://www.oed.com.ezproxy.library.yorku.ca/view/Entry/108303?rskey=ZwgyR0&result=3&isAdvanced=false#eid

Paradis-Peyton, T. (2010) Neoliberal Digitality, Labour and Leisure in an MMOG: An Ethnography and Analysis. Unpublished thesis, Carleton University. Ottawa, ON.

Riot Games Inc. (2009, February 25) **Official* League of Legends Teaser Trailer.* [Video file]. http://www.youtube.com/watch?v=RprbAMOPsH0. Retrieved May 2, 2012.

Riot Games Inc. (2011, April 1) *Summoner Showcase—#23.* [Video file]. http://www.youtube.com/watch?v=Qj1a02UYExY. Retrievd May 2, 2012.

Rose, N. (1999) *Powers of Freedom: Reframing Political Thought.* New York: Cambridge University Press.

Saussure, F. (1959) *Course in General Linguistics* (C. Bally and A. Sechahaye, Eds.; W. Baskin, Trans.). New York: Philosophical Library.

Suchman, A. (1987) *Plans and Situated Actions: The Problem of Human–Machine Communication.* New York: Cambridge University Press.

Stake, R.E. (1994) Case Studies. In N.K. Denzin and Y.S. Lincoln (Eds.), *Handbook of Qualitative Research* (pp. 236–247). New York: Sage.

Thoits, P.A. (1989) The Sociology of Emotions. *Annual Review of Sociology*, 15(1), 317–342.

Part III
Mediating Interpersonal Intimacy

8 Transconnective Space, Emotions, and Skype
The Transnational Emotional Practices of Mixed International Couples in the Republic of Ireland

Rebecca Chiyoko King-O'Riain

People use technology in a variety of ways to express their emotions on a daily basis. They text their mothers everyday with family news, they email their colleagues to share jokes and build camaraderie, they Facebook their school friends to exchange information and photos, and they use webcams to Skype their family and friends who live in distant places. In doing this, the type of emotion and the expression of emotion, as well as the emotion itself, are shaped in subtle ways by the technological forms being used and the technology in turn is shaped by its societal uses. Few studies have analyzed the relationship between emotions and the varied modes of technology used and yet everyday more and more people develop a wide-ranging repertoire of digital emotional practices.

Studies of Facebook show that 91 percent of teens on Facebook use it to connect not to strangers, but to people they already know and see frequently. Eighty-two percent use the sites to stay in touch with existing friends they rarely see in person (Lenhart and Madden 2007). Ellison et al. (2007) assert that Facebook is used to build 'maintained social capital' with people users previously knew, but that those who used Facebook most frequently used it to build 'bridging social capital' solidifying their weak ties into stronger ties due to the ease and low cost of staying in touch. Facebook also "lowered barriers to participation so that students who might otherwise shy away from initiating communication with or responding to others are encouraged do to so through Facebook's affordances" (Ellison et al. 2007: 1162).

While studies of Facebook tell us much about how people use this particular social networking site, they tell us less about what using it means in emotional terms to users. Building on studies of the sociology of emotion (Hochschild 1983), emotions can be understood as shaped by the social context in which they develop including in virtual or technological contexts. Unlike Facebook, this chapter shows that people are not just shaped by technology, but bend technology to their own emotional uses. While they may just display surface, or wide but not deep, emotions on social networking sites, they also use technology such as Skyping in mainly private (in their homes primarily), one-to-one, deeply emotional ways to maintain everyday emotional practices with pre-existing strong emotional networks. Particularly with increasing mobility, migration, and transnational practices, the emotions that bind people together are also shaped by the social conditions of migration and technological connections (Svasek 2008; Svasek and Skrbis 2007). Often people use technology in direct response to spatial and temporal distance to ease emotions such as loneliness or feeling 'friend' or 'home' sick. 'In using static technology such as email and Facebook, they mediate these emotions differently from face-to-face interactions. So while technologies like Facebook can solidify weak ties and even distant ties, it is very difficult to determine how deep those emotional ties really are. Outside of the 'status' update, the format of Facebook does not allow us to understand how these emotional ties expressed through Facebook are different from each other. In this sense, Facebook flattens emotional ties out so that connections to one's spouse may not appear much deeper emotionally than passing acquaintances on the face of it. Facebook then seems like a good way to maintain pre-existing relationships or as a "low maintenance way to keep tabs on distant acquaintances," but not to create deep emotional connections (Ellison et al. 2007: 1163).

Migrant global families, such as the ones studied here, have always sought ways to stay in touch with distant family members, local hometowns, and ethnic communities abroad. The use of technology to do so then is no new discovery. Migrants have always been on the cusp of finding and using technology to increase contact with home, find jobs, and emotionally stay in touch with those they love across the world. Burrell and Anderson (2008) found that Ghanaians living abroad used ICTs (Information and Communication Technologies) to 'look homeward,' but that they also used it "to explore the world more broadly searching for opportunities, information, contacts and new ideas" (p. 203). Baldassar (2007) found that the use of email in Italian Australian families increased the "frequency of transnational emotional interaction over time" which in the end increased the need for in-person visits to fulfill care obligations (p. 385). Mitra (2008) found that Indian migrants far from home use cyberspace itself to quell homesickness by blogging. Mitra states, "for people who have to move from place to place and undergo the diasporic experience, the anxieties of movement and placelessness produced by diaspora can be partly managed by entry into the cybernetic space produced by bloggers" (p. 457). The writing and reading of blogs helped to quell homesickness and ground migrants in a virtual if not their local community.

Ruckenstein (2010), studying the use of technology in Finland, claims that in effect, this use of technology and its link to mobility are often the prerequisites for better understanding the processes of globalization. "Mobility, both in the physical and computer-generated sense, is critical for these discussions, because it is understood as being at the heart of the . . . project of being a part of the global arena: mobility is valued over stability" (p. 502).

Unlike older types of technology, Skype has a media richness in its ability to deliver both audio and video information from a mobile platform. This can be a double-edged sword in terms of emotions. It allows a broader channel for emotions to be expressed across geographic distance and thus facilitates increased intimacy. However, such intimacy and emotional connection are also overwhelming for some users as it highlights the very fact of absence and distance while working to bridge them. Skype then gives some emotional stability within increasing experiences of mobility using technology to claw back some space/time for intimacy/connection to those you love and who love you from a distance. Technology, in this sense, could be seen as a "soothing force" from the hyper-mobile and distant and risky life of late-modern capitalism, not just as an intrusion into one's emotional and intimate life (Ruckenstein 2010: 504).

Technology in the past has been seen as an unnatural place for 'real' or 'true' emotions. The virtual world has been a space where true identities and emotions are masked or emotional companionship comes in the form of a techno-intimacy formed via virtual friendships (Allison 2006). Skype is different from these virtual examples because it blurs the boundaries of virtual and real life in new ways in an attempt to move away from inauthentic emotional presentation and practice to authentic emotional 'work'—to see, hear, feel, but not smell or touch those you love. Skype webcam use illustrates the lyrical movement between the actual and virtual as "part and parcel of the ordinary self" (Ruckenstein 2010: 509).

Technology, in this sense, is used more for "social" or "phatic" (Miller 2008: 387) reasons than just to exchange information. Anderson and Rainie (2012) found that the millennial generation both benefits and suffers due to their hyper-connected lifestyles. In their study of social networking, they found that nearly "20 million of the 225 million Twitter users follow 60 or more Twitter accounts; there are more than 800 million people now signed up for the social network Facebook spending 700 billion minutes using Facebook each month; Facebook users had uploaded more than 100 billion photos by mid-2011 and YouTube users upload 60 hours of video per minute and they triggered more than 1 trillion playbacks in 2011—roughly 140 video views per person on earth" (2012: 9). With all this technology connecting people to their friends and those they love, these techno users are hyper-connected. They are good at multitasking with a level of quick thinking and multi-thinking, but are also compulsive, impatient, and have less 'deep' or critical thinking. Does this mean that there are fewer 'deep' emotional connections for this generation?

Valentine (2006) argues that the use of the Internet "stretches intimacy beyond the boundaries of the domestic" (p. 387) and that there is an increasing belief that intimate relationships are changing for the worse (Beck-Gernsheim 2007). However, the actual technological practices of intimate emotions "may not be changing as much as anticipated because the Internet offers a way of rearticulating offline practices of everyday familial life that can, at least in part, compensate for the limits of intimacy elsewhere" (Valentine 2006: 387). Geographical and temporal closeness then may not be necessary for emotional closeness. However, Valentine is careful to point out that there are limits and that ICTs are not substitutes for physical intimacy and "the absence of actual touch can serve only to accentuate the emotional pain of missing or longing for another body" (p. 388).

Alinejad (2011) who studied online blogs of Iranian immigrants also found that these transnational spaces illustrated 'transnational embodiment'—a strong awareness of bodily presence in offline locations and situations, which continually informed and shaped online expressions.

> The importance of physical travel to, proximity to, and sensory impressions of particular places within two bounded, politically distinct nation-states shows that diasporas rely heavily on embodied experience in constructing transnational spaces not only on psychic ties and recalled memories. Members of second-generation Iranian diaspora reveal unique types of embodied ties to a diaspora 'home' through their apparent search for authenticity. (Alinejad 2011: 43)

These are emotions, which traverse along "intimate circuits" (Bloch 2011: 13) but are not disconnected from the sights, smells, and bodies of real places. Part of how global capital shapes people's lives involves looking beyond just the working conditions of people's lives to see the emotional and intimate worlds of migrants (Bloch 2011: 14) and how they are shaped.

A more emotionally fulfilling way to communicate with a facial and bodily dimension with those you love increasingly is to use Broadband Based Visual Communication technologies. There have been studies of Broadband Visual Communications (BVCs) use in order to share information and build social ties (O'Donnell et al. 2010). The focus of this research was on video conferencing and its use as a form of synchronous audio-visual communication. They find that BVC technologies allow for "the exchange of visual information like facial expressions that encourage trust and intimacy," (p. 529) and are also cheap, environmentally friendly, and a good way for people who are far apart to connect. And while Hardey (2004) argues that these types of technologies "are supplementing or replacing traditional routes to potentially romantic encounters" (p. 207), I was interested in finding out how the regular use of Skype as a BVC impacted the emotions, emotion practices, and expressions of those who use them to communicate across great distances in non-technologically established mixed international couples. By 'mixed' I mean that one partner was from Ireland and one was not originally, and by international I mean that the couple defined themselves as international or global in focus and were in regular contact with loved ones (partners and others) in distant (often culturally diverse) locations. But why study this topic in Ireland?

IRELAND

From 2000–2010, the Republic of Ireland was one of the fastest growing places in Europe in both economic and demographic terms. The Celtic Tiger economy brought many multinational companies into Ireland and the demand for workers in many industries was satisfied by non-Irish migrants. This has meant that Ireland has been one of the fastest changing societies in Europe in terms of ethnic/racial diversity in the last 10 years. Today 10% of the population in Ireland is non-Irish born and it has the highest birth rate in Europe—16.8 children born per 1,000 inhabitants in 2009, compared to an EU average of 10.7 (Taylor 2010).

Alongside increasing growth and cultural diversity, Ireland also had one of the most recent expansions of broadband access in Europe. In 2005, 45% of households with at least one person ages 16–74 had a computer connected to the Internet. By 2008, that number rose to 62% and broadband use had increased from 7% in 2005 to 43% in 2008 (CSO 2008). More people had computers and access to the Internet at faster speeds needed for VoIP (Voice Over Internet Protocol) or Broadband Visual Communication (BVC) technologies.

This chapter brings together a study of emotion with the impact of technology by examining how international mixed couples and their loved ones 'do' or practice love in relation to the use of Skype BVCs. The focus of the study is on *transconnectivity*, which I term to be about the practices that people 'do' to create and maintain emotional connections, both through technological and emotional means, that people have across nation-state, cultural, and political borders. It is this criss-crossing of places and spaces through emotional networks on technological platforms which enables transnational senses of belonging, habituses, and identifications. The chapter asks: How do mixed couples use technology, particularly new broadband visual communication technology such as Skype, to maintain, create, and sometimes cut off emotion networks? What effect does the technology have on the ways they experience emotions and their understanding of space in the global world?

METHODS

The data in this article come from an in-depth qualitative interview study of mixed international couples living in the Republic of Ireland. I conducted 40 interviews in English in 2010–2012 with same sex and heterosexual couples (ages 26–60) and adult children, from Ireland, France, Canada, the United States, the United Kingdom, Malaysia, India, Sri Lanka, Poland, Zimbabwe, and China living in Ireland, the United Kingdom and the United States. In Ireland, interviewees were from: Cork, Kildare, Galway, Tipperary, Dublin, and the surrounds. The interviews were digitally recorded and then transcribed.

Thirty-five of the couples interviewed were heterosexual and four were same-sex couples (three lesbian and one gay male couple) and one refused to state. All couples had one international partner or spouse who described himself or herself as 'international' and not Irish. More non-Irish women with Irish men were interviewed and constituted about two-thirds of the interviewees. There were men interviewed, but again they tended to be in relationships with non-Irish women. While these interviews, due to sampling methods, cannot be generalized to all intermarried couples in Ireland, they do mirror demographic and gendered trends in intermarriage in Ireland.

The chapter concludes that the mixed international couples interviewed here used Skype to create a *transconnective* space. To do this I found that: (1) Skype, and technological devices more generally, were often introduced or learned with the explicit goal of maintaining and creating emotional ties across generations and partners in international families; (2) Skype enabled interviewees to maintain distant geographic but emotionally close ties over time and space; and (3) the accentuation of pain of missing 'actual bodies' (Valentine 2006) actually made some transconnective relationships shift away from Skype precisely because of its bodily and facial intensity.

EMOTIONS DRIVING TECHNOLOGY—GETTING ON SKYPE

Many have written about how technology, and in particular digital social networking technology, is making the world a smaller place (Miller 2011). It is easier now to reach out across the globe to meet and sometimes become intimate emotional partners with those who are from very different cultural backgrounds, religions, languages, and emotional understandings than oneself (Ben-Ze'ev 2004). Technology, and in particular, mobile technologies, should be making the global world a more connected place where the flow of emotional connections can take place in almost any space (Elliott and Urry 2010). However, some social networking sites (Miller 2011) and online dating (Ben-Ze'ev 2004) move primarily from online to offline relationships where the online content is fairly superficial and when people want to have more emotionally fulfilling interactions they move the relationship offline. Skype is slightly different to these because of its ability to enrich face-to-face pre-established offline relationships when participants can't physically be together in the same time and space. The *emotional intentionality* of Skype then is very different as it is focused on close emotionally connected relations and not old school chums or passing acquaintances.

The distance created by migration of one partner in my sample, sometimes motivated older relatives, such as grandparents, to become more technologically proficient and to join Skype, driven primarily by the desire to maintain strong emotional connections to their grandchildren. Maria, from Milan, Italy, met her husband Liam, from Dublin, Ireland, in Italy while studying. She later married him and they migrated to Ireland, where they live with their 5-year-old son on the outskirts of Dublin. Her parents, back in Milan, treat Skype as a form of a phone call, but as she and her husband talk more about how they use Skype it is clear that it moves far beyond a voice-only phone call. They explain:

Maria: We do schedule it [Skype call] because for my parents it is a bit of a procedure. They are not really spontaneous users of the Internet. So you kind of have to say, "oh we will call . . . ." But then we are kind of flexible, because obviously the main aim is to see the grandchild, so he runs around and he shows them things and we kind of run around with the laptop. It is kind of nice.

Liam: Yes so I wouldn't say it is like a phone call in that way, I mean they bake and stuff and talking to the computer, and sometimes if he is not in the mood he might be watching TV in the evening, if he is tired or something and they would be watching him watching TV.

Maria: Yes it is very flexible. Or sometimes I am like, "oh you play with your grandparents and I am going for a shower". And he is just there at the computer playing games with my parents.

It turns out that both parties have Skype now on mobile/laptop devices and try to use the mobility of the technology to 'show' more of their lives. In fact, while they only Skype once or twice a week, they often leave it on so they are capturing the ebb and flow of daily life doing things like baking or even watching their grandson watch TV via Skype. They argue that while they feel trapped behind the computer in order to see him, they enjoy just seeing their grandchild in his natural setting and they feel they are emotionally closer to him because they chat with him in Italian in a more natural way. This type of unscheduled intimacy or everyday emotional connection was not possible for them via email or the telephone, but the Skype webcam allows them not strictly to interact with each other face-to-face, but also to watch each other, thus mediating emotions in a more casual and natural way—almost like 'being there.'

Malia a Turkish/French woman married to an Irish man living in west Dublin, discusses the difference between using the phone and using the visual cues of Skype to communicate with her parents now living in France. Malia explains:

There is a difference. I see my daughter. She changes her whole behaviour as well. She jumps. She starts to get excited and she jumps on the couch and she is a bit more, I don't know how you say it in English, she is showing them that she loves them whereas on the phone she would be distracted by the TV or other things. But she really enjoys talking to them and she'd be chatting away in English sometimes and they don't understand her. Sometimes she would be going on in English and they would say, 'talk in Turkish.' She says, 'I am talking in Turkish.' But she doesn't realise that she isn't. With the webcam on Skype they are closer.

Malia explained that when her daughter was younger, it was important for her parents to see the baby almost weekly because she was physically changing so quickly and couldn't speak on the phone. The visual aspect, while not completely bodily because they can't hug, does come into focus more clearly with Skype and changes her daughter's behavior. Her attention to the visual, as perhaps a more general increasing importance of visual stimulation in her generation, is important and changes her behavior. She is able to show them her ballet dancing, sing for them and point out her new toys and books. She is also entertained for longer periods of time and thus interacting with her grandparents more deeply because of the visual cues. For Malia, she argues that being able to 'see' her grandparents makes them more real to her daughter and reminds them (because young children have a shorter memory than adults) of who her grandparents are and that this made their geographic distance less emotionally painful. Malia also told me that she found it interesting that her daughter can understand Turkish even though she doesn't speak back to her parents in Turkish and that this is primarily because of seeing and speaking with her grandparents on Skype since she is not really exposed to the language in her daily life. Both Maria's and Malia's parents don't speak English and Skype has meant that they use technology to make emotional but also important cultural and linguistic connections to their grandchildren. Without the grandchildren learning the language of the grandparents, there would be no means or mode of communication and hence no emotional relationship which would be common to all three generations.

GEOGRAPHICALLY FARTHER, BUT EMOTIONALLY CLOSER

Veronique, originally from Marseille, France, but now living in County Limerick, also illustrates how her use of Skype to her family in France helps her Irish/French children to maintain not only their emotional ties to people, but to the French language and place. She says:

> I'd be on Skype everyday. It would be on and if somebody wants to ring me or somebody wants to ring somebody that would be on. I use it for work as well, like with the university, I am still with the French University and I am using it to talk away with people over there. The phone in Ireland is very expensive. My children talk to my parents. I would say it helps them maintain a relationship with them. My husband is older so they don't have grandparents in Ireland and his brother and sister; they are kind of non-existent. The only family they would have, even though there are some here, would be my family in France, if you see

what I mean. By using Skype, even though my family is further away distance wise, they are closer emotionally to them.

It is clear that the local Irish relatives are not the primary source of emotional support, but that using Skype allows her children to speak with and have a strong emotional connection to her family in France and to the French language, which they use exclusively on Skype. Veronique talked at length about the ability to speak French perfectly, with the correct accent, and how that would allow them to be seen in France as 'real' French children even though they are half Irish and live in Ireland. Her goal for her children was to keep up strong emotional attachment to her family, but also to her hometown and the French language through the frequent use of Skype.

While there are clearly similarities in terms of language and cultural connections being fostered by these French grandparents of their mixed French/Irish grandchildren, the children themselves identify much more strongly with being French than being Irish even though they live in Ireland and speak Irish as well. For Veronique, she has made a conscious choice for her children, to connect them both physically, spending the summers in France, but also via Skype culturally and linguistically with France and her side of the family. Even though her Irish in-laws are in the same village and same school as her children, she describes how the Irish cousins would have almost no relation to her children, not inviting them to birthday parties over other classmates. Who you choose to Skype in this instance may not be your local neighbor or relations, but instead, those you love and who love you from a distance. They are the emotional connections that Veronique makes the effort to use Skype for when she is keeping it on all day. She would never dream of Skyping someone locally, but uses it daily to connect to France, French speakers, and her French family as if they were around the corner.

TRANSCONNECTIVITY OR DISCONNECT?

However, not all of the emotions expressed, created, or maintained on Skype were positive. There were some emotionally negative experiences described in the interviews. Lily, who is a self-described rebel, originally from Seattle, Washington, is married to her Irish husband and lives with him and their four children in the west of Ireland. Lily loves to knit and described herself as "raising kids full time" and her husband as being "unemployed for quite a while." Lily left the United States spontaneously and didn't inform her parents until the last minute. They were shocked and hurt by her abrupt departure with their three grandchildren to Ireland and that rift took time to heal. She describes here how she has used Skype and how it didn't always create emotional closeness.

> Skype can help in a good way but it can also help in a bad way. Like my mum is a bit of an alcoholic so she Skypes at night our time, so she

Skypes at midnight [unclear 57 44 15] and she is not making the best impression on her grandkids. They haven't seen her in four years and when she calls on Skype she is completely drunk and slurring and the kids are like, "oh my God, your mother, look at her." They don't even call her their grandmother; they are like, "your mother." Your grandmother too. And then it is a bit embarrassing. But my dad is different. He is short, sweet and to the point. A long conversation for him is 20 minutes but that hardly ever happens. We have ours [Skype] on all the time. It is on 18 hours a day so anybody can call us whenever they want but out of my dad, my mum and my sister, none of them are signed in all the time.

Clearly for Lily, fraught emotional relationships are not resolved or made stronger by the use of Skype. In some sense, by leaving the Skype on all day, she is open to communicating with her family if they are signed on, but in reality, her family life is quite busy and she appreciates the frequent, but shorter contacts.

Others also described their frustrations with the emotional and bodily limitations of Skype. Isabelle, a French woman from Paris, France, now living in the greater Dublin area with her long-term Irish boyfriend, described how they use Skype when they are traveling apart from each other and the limitations of that over the long term. Isabelle says:

Isabelle: When we are apart, we talk every day, we have Skype phones.
Interviewer: And do you think it has helped you to . . . ?
Isabelle: Yes, it helps to a certain point. Like at this point I am sick of it. We call each other because we got into that routine but I told him the other day that I am really sick of talking on the phone. I need to just be in the same place and do something together. But yes we made great use of it, the Skype phone and the video Skype.

A final interviewee explains how he sometimes prefers the telephone to BVCs like Skype because he feels that Skype is almost too close emotionally. Balaji, a multiethnic Indian/Irish college student in his early 20s from Dublin's Southside explained that his parents divorced when he was quite young and his Irish father emigrated to Australia while his Indian mother and sister (who has special needs) remained in Ireland. Balaji states:

> The time difference is hard but not only that, it is the time that I am working and then in college and we have to arrange a time when he is not working. And I think it is also easier to use the phone, not on practical levels but on emotional levels it is easier to use the phone.

Interviewer: That is interesting, why?
Balaji: I don't know, it is just hard to say goodbye when you are looking at the person, it is very difficult to disengage or disconnect and I hate the red button that says, 'disconnect call.' I hate pressing that. And he is always quite hesitant; it is easier just to use the phone.

Balaji was quite clear that although Skype technology helps them to close the geographic space between Australia and Ireland, the time differences and the time rhythm of his day and his father's day make it difficult to coordinate communication. He also spoke of the bulky nature of being dependent (in his case) on the laptop or computer to communicate when sometimes he just wanted to be able to text or use a mobile device (like a phone) to have a quick chat or send a quick message and he did talk about working to try to afford a smart phone (iPhone) so that he would be able to do this. His main objection, though, to Skype was emotional. He found it just too hard to be staring his father in the eyes and have to punch the red button to disconnect the call—hence making his image vanish into the ethernet.

CONCLUSION

This chapter has explored how mixed international couples, and in some cases their children, use broadband visual communication technologies such as Skype to maintain and sometimes strengthen emotional ties to distant people, places, and cultures/languages. While it is clear that Skype webcam technology is helping some mixed migrant families to stay in touch, keep up language acquisition, see physical development of children, etc., most often the explanation given for using (or learning to use) the technology is an emotional one. The technology affords emotional connections to stay strong over long distances and time separations apart. While Skype is not a first choice, it is a good second choice—the next best thing to being together in person, and while apart, Skype helps people to manage the desire to be together again.

Skype then is changing how emotions are expressed and how they continue to develop with international families. They have become a key tool to maintain emotions between generations of migrant families and to experience 'everyday' as well as scheduled chats with distant family members between 'real' visits, which involve real and costly air travel. All the interviewees discussed how Skype had changed their emotional lives and expressions because of the cheap and easy-to-use nature of Skype—some even described it as revolutionizing their children's relationships to their parents—both as grandparents learned to use Skype and technology, but also as children learned to present themselves and their lives to their grandparents via Skype.

Skype, however, clearly had its emotional limitations and was not a replacement for a hug, bodily contact/intimacy, or even just a chat down in the local pub over a cup of tea. Balaji, Isabelle, and Lily make it quite clear that while they use Skype, there are emotional negatives or limits that come with it such as bad emotions as well as good ones fostered by the use of the webcam and limits on how much time they can spend on Skype or how bulky and immobile the technology can be (although this is changing with the increased use of tablet and mobile devices with Skype webcam capabilities). One of the strongest reasons, though, not to use Skype, was emotional. They felt that it was just too difficult emotionally, to face their loved ones and then have to disconnect.

REFERENCES

Alinejad, D. (2011) Mapping Homelands Through Virtual Spaces: Transnational Embodiment and Iranian Diaspora Bloggers. *Global Networks,* 11(1), 43–62.

Allison, A. (2006) *Millennial Monsters: Japanese Toys and the Global Imagination.* Berkeley, CA: University of California Press.

Anderson, J.Q. and Rainie, L. (2012) Millennials Will Benefit *and* Suffer Due to Their Hyperconnected Lives. Retrieved from http://www.pewinternet.org/~/media//Files/Reports/2012/PIP_Future_of_Internet_2012_Young_brains_PDF.pdf on 21 July 2013.

Baldassar, L. (2007) Transnational Families and the Provision of Moral and Emotional Support: The Relationship between Truth and Distance. *Identities: Global Studies in Culture and Power,*14, 385–409.

Beck-Gernsheim, E. (2007) Transnational Lives, Transnational Marriages: A Review of the Evidence From Migrant Communities in Europe. *Global Networks,* 7(3), 271–288.

Ben-Ze'ev, A. (2004) *Love On Line: Emotions on the Internet.* New York: Cambridge University Press.

Bloch, A. (2011) Intimate Circuits: Modernity, Migration and Marriage Among Post-Soviet Women in Turkey, *Global Networks,* 11(4), 502-521. doi: 10.1111/j.1471-0374.2010.00303.x

Burrell, J. and Anderson, K. (2008) "I Have Great Desires to Look Beyond My World": Trajectories of Information and Communication Technology Use Among Ghanaians Living Abroad. *New Media & Society,* 10(2), 203–224.

Central Statistics Office (CSO). (2008, December 10). Information Society Statistics First Results 2008. Cork, Ireland: CSO.

Elliott, A. and Urry, J. (2010) *Mobile Lives.* London: Routledge.

Ellison, N.B., Steinfield, C. and Lampe, C. (2007) The Benefits of Facebook "Friends": Social Capital and College Students' Use of Online Social Networking Sites. *Journal of Computer-Mediated Communication,* 12, 1143–1168.

Hardey, M. (2004) Mediated Relationships. *Information, Communication and Society,* 7(2), 207–222.

Hochschild, A.R. (1983) *The Managed Heart: Commercialization of Human Feeling.* Berkeley, CA: University of California Press.

Lenhart, A., Madden, M., Macgill, A, and Smith, A. (2007) Social Networking Websites and Teens. *Pew Internet & American Life Project.* Retrieved from http://www.pewinternet.org on 21 July 2013.

Miller, D. (2011) *Tales From Facebook.* Cambridge, UK: Polity Press.

Miller, V. (2008) New Media, Networking and Phatic Culture. *Convergence: The International Journal of Research into New Media Technologies,* 14(4), 387–400.

Mitra, A. (2008) Using Blogs to Create Cybernetic Space: Examples From People of Indian Origin. *Convergence: The International Journal of Research into New Media Technologies,* 14(4), 457–472.

O'Donnell, S., Molyneaux, H. and Gibson, K. (2010) A Framework for Analyzing Social Interaction Using Broadband Visual Communication Technologies. In T. Dumova and R. Fiordo (Eds.), *Handbook of Research on Social Interaction Technologies and Collaboration Software: Concepts and Trends* (pp. 528–541). Hershey, PA: IGI Global.

Ruckenstein, M. (2010) Toying With the World: Children, Virtual Pets and the Value of Mobility. *Childhood,* 17(4), 500–513.

Svasek, M. (2008) Who Cares? Families and Feelings in Movement. *Journal of Intercultural Studies,* 29(3), 213–230.

Svasek, M. and Skrbis, Z. (2007) Passions and Powers: Emotions and Globalization. *Identities: Global Studies in Culture and Power,* 14, 367–383.

Taylor, C. (2010, July 27) Ireland Has EU's Highest Birth Rate. *Irish Times.* Retrieved from http://www.irishtimes.com/newspaper/breaking/2010/0727/breaking38.html on 10 August 2012.

Valentine, G. (2006) Globalizing Intimacy: The Role of Information and Communication Technologies in Maintaining and Creating Relationships. *Women's Studies Quarterly,* 34(1 & 2), 365–393.

9 Send Me a Message and I'll Call You Back
The Late Modern Webbing of Everyday Love Life

Natàlia Cantó-Milà, Francesc Núñez, and Swen Seebach

INTRODUCTION

New communication technologies are playing an increasingly important role in the quotidian webbing of our lives. This is especially true for love relationships. Lovers use their mobile phones, email accounts, and webcams to communicate with each other in the most diverse situations and circumstances, and for the most varied reasons: shopping lists, the nanny's phone number, love poems, or the most intimate confessions.

The possibilities of remaining in touch in spite of distance, or the possibilities of creating 'artificial' distance by chatting from the bedroom to the living room, of carrying on arguing in the tube on the way to work, or of discussing the latest movie seen together while sitting at the desk at work are webbing new forms of living and experiencing partnership, and subsequently enabling new possibilities of defining, experiencing, and expressing commitment.

To be always there for each other now does not only imply the more-or-less vague promise to stay together, it may mean being there for each other at every minute, every second, 24 hours a day, wherever one is, for there is always the possibility of 'being there,' of 'being in touch.' At the same time, despite living together, couples might learn to perceive—and communicate with—each other increasingly through written words and digitalized images, in the same way we learn to perceive our colleagues or old school friends via, say, Facebook.

In this chapter we present the results of the analyses of 40 autobiographical interviews, mainly collected in two different European cities: Barcelona and Berlin. Our interviewees narrated their love stories; their daily practices; their conflicts, hopes, and problems. All these results are seen through the particular lens of this work, which focuses upon the ways in which contemporary lovers use electronic communication, and the ways in which these uses are embedded in the whole relationship: sometimes making things possible which otherwise would not be possible in the same way or as easily and immediately, sometimes accelerating events, sometimes allowing possibilities (and sometimes enabling responsibilities), of multi-presence and forms of obligation, which shape the relationship.

THEORETICAL FRAMEWORK

Over the last decade the new possibilities and realities of communication through new communication technologies (above all the Internet and mobile phones) have caught the attention of a great number of social scientists of many kinds.[1] Relationships that began on the Internet, or long-distance relationships that have been sustained and made possible through the use of electronic communication, have become a new and highly sought-after research field in the social sciences—and this is surely for good reasons.

However, as interesting as the analyses of these highlighted phenomena are, the analysis of the slow penetration of the new technologies of communication into the love relationships of people who live together or close to each other, and who do not perceive these new forms of communication as constituting an important or central part of their relationship, is also highly relevant. Within these relationships, the people involved hardly reflect on their uses of electronic communication on a daily basis. This lack of reflection does not mean that electronic communication is not being made use of. To the contrary, as our interviews have shown, it is being used, even for the very organizing of everyday life and quotidian responsibilities, and often also for expressing feelings, confessions, and communicating wishes and desires, of which some would be felt to be too embarrassing to communicate face-to-face. New forms of communicating (via the use of electronic devices) have penetrated the everyday of these relationships so subtly that the people involved do not realize it on a daily basis, and do not reflect upon the increasingly important and central role that these forms of communication attain in the everyday life webbing of their love relationship, and thus of their love life.

Thus, despite seldom being perceived as such, in their 'taken-for-grantedness,' these new channels of communication between lovers slowly reshape their patterns of relating with each other; their patterns, forms, and styles of communication; their perspectives on each other; and their performances for each other, thus creating new expectations, hopes, and desires, erasing older ones, superimposing or supplementing them, and so modifying the bonds webbed in couples.[2]

In our analysis of the bonds webbed between couples in quotidian life we took Simmel's ideas on reciprocal actions and effects (*Wechselwirkung*) into consideration (Simmel 1999: 70). We furthermore worked with the idea that certain forms of interrelating with each other change according to the form and the material by which the bond is webbed. As Simmel mentioned in the chapter The Style of Life in *The Philosophy of Money*, intellect and money (as forms of sociation) have both helped to shape social relations in modern society, by levelling out individual characters and social relations, crystallizing social processes and converting them into objects of value and knowledge (Simmel 1989: 436). Interrelated with these processes, however, a stimulation of emotionality and romanticism takes place.[3] This chain of thoughts led us to question: what happens when social processes are increasingly based on a new medium?

Eva Illouz has touched upon this question. Her analyses of love relationships from the beginning of the 20th century until the 1990s (Illouz 1997) and her analysis of love relationships webbed via the Internet at the end of the 20th century (Illouz 2007) are very helpful for understanding some of the crucial differences between love relationships in Simmel's time and in the times of electronic communication.

Apart from these developments, Illouz's analyses of romantic love relationships indicate a crucial difference between everyday life practices and extraordinary moments in the webbing of love relationships, a difference which, Illouz claims, is the basis for the functioning of love relationships in modernity, and which makes love relationships a crucial dimension for the webbing of modern society in general. Illouz furthermore points to emotions as being crucial elements within the process of webbing social bonds, formed and performed in love relationships; they are the material out of which desires and expectations are webbed.

However, in her analysis of love in the late-20th century, Illouz does not focus on the distinction between everyday life practices and special moments in love relationships, but instead focuses her analysis on the emerging practice of Internet dating. We decided to follow the way pointed to in *Consuming the Romantic Utopia* (1997), focusing our analysis on the dynamics and meanings of everyday life practices and special moments that can be webbed between couples in times of electronic communication: apart from everyday life practices, love relationships find their crucial meaning in moments of 'liminality' as Victor Turner (1979) calls them, moments in which the everyday order and rhythm is suspended and the couple enter a space of enchantment. Liminal moments are crucial for the webbing of love relationships, as the memory of these liminal moments, and a sort of ritualized remembrance of them, is crucial for the renewal of the commitment to the love relationship. In the discussion of our results we shall focus upon the liminal (or liminoid) moments in the webbing of love relationships, online and offline.

THE INTERVIEWS

In order to undertake this research 40 interviews were carried out (mainly in Barcelona and Berlin) during the year 2011. In many interviews the issue of electronic means of communication was thematized without the interviewer needing to explicitly ask about it.[4] In these cases the last part of the interview served to explore in greater depth issues that had already been mentioned, rather than bringing to light issues that had remained undiscussed. However, there are also interviews in which the question regarding electronic communication led to anecdotes and stories arising, which were of great importance for the relationship, and which had not been mentioned before.

Furthermore, when our interviewees reported the ways in which they electronically communicated with their partners, they referred to a great variety of forms of electronic communication (instant messaging, emails, Skype phone calls, Skype video conferencing, mobile texting, mobile phoning). Despite the fact that not all interviewees made use of all the previously-mentioned electronic communication media, they all specified that they did not make the same use of all the electronic media that they were users of. Moreover, they told us that they considered certain means of communication to be appropriate for expressing and discussing certain issues, but inappropriate for others. The interviewees did not all coincide in the judgments they made about the correctness of the use of certain electronic media and not others for, say, organizing daily tasks, expressing deepest feelings, confessions, or erotic games within their love relationship. But they did all agree on the fact that they had diversified and specialized their uses of electronic communication regarding the media used. The specialization of the uses given within the couple to different forms (and devices) of electronic communication depended basically on three factors: (1) distance; (2) budget; and (3) preferences, personal abilities, and (pre)judgments about the worth and qualities of one or the other form of electronic communication.

On Distance

In those cases in which the couple lived nearby or together, the uses of mobile phones were much more significant for the couple's everyday life than the uses of computers. This changed radically in cases in which couples lived far from each other, or spent time apart from one another. In these cases, computers played a crucial role.

ON BUDGET

Closely connected with the previous point, interviewees engaged in long-distance relationships reported that they made particular use of the computer and different computer applications in order to interact with their partners. One interviewee referred to Skype as the "love provider" of the relationship, the place "where they meet up" (Mark, 29). He furthermore explicitly stated that they made use of Skype because it was free, and asserted that "Without Skype our relationship would be much more expensive (laugh)" (Mark, 29).

Short instant messages play the role of a mobile text in long-distance relationships, or moments apart (abroad). We also saw a tendency (above all among the younger generations) to give up mobile texting for instant messaging applications for the mobile phone. In all cases, the argument for this change was monetary (instant messaging applications are free).

On Preferences, Personal Abilities, and (Pre)judgments

Regarding preferences and assessments of the written word (above those means which allow oral and visual communication), our interviewees were divided. Fourteen interviewees referred more or less explicitly to the written word as being "useful" in order to either clarify their own thoughts or to get a message through when there were problems in the relationship. Five people reported that they had the custom of sending long emails to their partners in times of crisis. Seven asserted that many times it was the best way to get a message through because a written message could not be interrupted, and three others presented the use of the written word as a means of cooling down the argument. In these cases we realized that there was a clear difference in relation to those people who were better able to express themselves through writing, and those who found it very difficult. The ability to write, and to feel comfortable with the written word, created a divide between the possible usages made of electronic forms of communication in order to interact with the partner.

Twelve interviewees explicitly mentioned the fact that they tended to argue more with their partners when they were communicating electronically than when they were together. Furthermore, one interviewee, who was engaged in a long-distance relationship, added that Skype was a way of palliating this effect, because one could at least see the other person's facial expression and most immediate reactions to the things said and done. Four interviewees highlighted the fact that they had explicitly agreed not to discuss difficult issues with their partners via electronic communication, only face-to-face. Furthermore, 10 people simply reported that they did not use electronic communication for carrying on with arguments (despite not mentioning any explicit rule behind this behavior).

> By email or text? No never. . . . You get confused by the words . . . if you cannot see the other person's reaction. Perhaps he means something as a joke, and you take it as something serious. At the beginning we had long arguments, until we decided to establish the Blackberry rule. No arguments until we see each other. (Andrea, 22)

Sixteen interviewees reported that they liked to exchange short mobile texts in order to seduce, re-enchant or make their partners laugh when they were apart (in relationships that implied cohabitation or vicinity). For instance:

> A message is like a pill, something very concrete that you transmit to the other person, it is like a poem (. . .) but it is . . . it depends on what you say, you say something and that's it; in contrast, a conversation implies that perhaps one wants to talk, and the other doesn't, or just cannot because I am learning, working, . . . wherever, I couldn't reply . . . and why don't you reply? Instead a text is just that . . . and if you read it, and it is beautiful, it reaches you more. (Pau, 28)

Beyond the specialization of means of electronic communication, and the selection of the uses made of electronic communication by each couple, one important result of our research is the finding that, when lovers communicate electronically—be it through Internet or mobile phone (or a combination of both)—certain ways of expressing emotions, and also certain emotions, are generated in ways that differ from those which may have occurred if their communication (and thus relation) had taken place otherwise. Above all, we find an acceleration in the tempo of communication, and an increase in the expectations of response. And this applies to everyday life organizational matters as well as to the most romanticized waiting for the first date.

DISCUSSION

Let us now differentiate the most important issues that are dealt with electronically between couples (tasks, desires, wishes, disputes, feelings) and the emotional implications they have, after which we will analyze the consequences of dealing with these issues electronically.

On Daily Tasks

Electronic communication plays a crucial role in the organizing of daily tasks, and especially for those couples who cohabit to a certain extent. In many cases, couples do not organize their day in the morning before they leave home, but during the day, by means of their mobiles.

In the case of older relationships with cohabitation ('older' referring here to the age of the relationship, not the age of the people engaged in the relationship) in which we found that electronic communication does not cover all the couple's communicational needs, it is used above all for organizing their daily tasks. The interviewees in this life situation did not tend to value the importance of this means of communication much in their narrations. However, they did describe situations in which electronic communication unexpectedly failed, and this failure led to great frustration, anger, and the impossibility of going on with their daily tasks as they had planned them—or, more likely, taken them for granted.

This does, of course, not mean that people who have been living together for a longer period of time do not use electronic communication in order to express their feelings, their sexual longing or desires, or to communicate their innermost dreams. The point is that they do it less often, in general terms, than the rest of interviewees that we talked to. Moreover, it should be stressed that we did not find a single person who was engaged in a cohabitation relationship who did not use electronic communication in order to organize their everyday life with their partners, even if it was only one or two mobile phone calls a day to ask for something to be done.

We found people who, after some time in the same relationship, reduced their use of electronic means of communication in order to interact with their partner—in terms of frequency of use, and of the media used to communicate. Thus, for instance, in the initial phase of their relationships, they had used mobile texting quite often to briefly comment upon the beauty of the evening spent together, or the love they felt for each other. Then after a while spent living together, they stopped using mobile texting, and only used their mobiles to call each other at lunch time in order to check on how their partner's day was going, and to organize their evening.

> Yes, at the beginning we did some texting. After going out, for instance, to say that the evening had been so nice, yes . . . for instance. Now we just do some phoning—yes, at lunch time, and when we need something from each other. (Beate, 40)

In the case of long-distance or partly long-distance relationships we realized that the organizing of daily tasks was not a big issue. Notwithstanding, we found out (and this we were not expecting, to this extent) that these couples had an idealized imagery of what sharing daily life with their beloved ones would be like, and they spoke about it with a certain melancholic tone:

> I know his innermost wishes, his sexual desires, I have seen him exposed in ways nobody else ever has. . . . But I do not know if he drinks his coffee with milk and sugar. . . . I don't know what kind of socks he prefers to wear. . . . I know nothing of these details. . . . It is strange. (Eva, 50)

Two interviewees engaged in long-distance and partly long-distance relationships told us that they used Skype to try to simulate a sort of daily life together. And despite knowing that they were in some way simulating, they presented these moments as being really dear to them, and confessed that it was one of the most crucial issues when sharing lives apart from one another. Simulating cohabitation, through the development of everyday life routines that they would naturally share if they were physically together, "gives perspective to the relationship." Thus, they reported that they used to cook the same dish together, using their laptops with Skype webcam chat open in their kitchens, and thus they had supper together and shared a bottle of wine and a nice meal over the distance (Lisa, 25).

Another person told us that he and his lover had computer screens and computers connected to Skype 24 hours a day in a room at their respective homes, and they would just run into each other as if they were "just doing (their) own individual things, and paying attention to (their) work, and then running into each other in the living room for a nice cup of tea or coffee together, and a brief chat, just as you would do if you run into each other when you are in the same apartment, do you know what I mean?" (Ian, 38).

These findings point to the importance that quotidian life has for a relationship—whether it be a cohabitation relationship or a long-distance relationship. For those couples engaged in a long-distance relationship, the quotidian becomes something special, something that they cannot have on a daily basis. The creation, imagination, and performance of quotidian life is turned into an important ritual that webs the relationship together, like a visit to a restaurant or a romantic email does for relationships with cohabitation.

On Emotions

We found a more intense use of electronic communication to express emotions in the case of long-distance and partly long-distance relationships, as well as in the cases of relationships with cohabitation among the younger or better-educated couples.

In a direct or indirect way, they all coincided in stating that emotions are expressed differently on the Internet and through mobile phones. They also coincided in affirming that Skype video calls, instant messages, or emails differ in the possibilities they allow for emotional communication. They all have their advantages and disadvantages: long emails allow for reflection and distance, to be able to think things over before sending them so as to get the right message through and, if possible, achieve the desired effect; instant messages and mobile texts allow for brief enchanting moments, and also the immediate communication of pragmatic and organizational pieces of information; Skype video calls enable visual contact with the other person, to engage with the game of overcoming the distance, to enchant the other and oneself with the reflected images on the screen.

However, what our interviewees stress most is that none of these forms of electronic communication allows the same empathy and immediacy as face-to-face communication. The same dimension of "feeling the other person, knowing what she needs, how her hands tremble" (Peter, 27).

The knowledge (and experience) that these means of communication do not allow the nuances of emotional communication that are highly necessary for a successful communication between lovers makes the users become more inventive in order to overcome these communicational shortages. The resources for building these bridges toward the other and his/her emotions vary: from the use of emoticons to the generation of private rituals of intimacy, they try to overcome the rigidness and coldness of the means used and they try to domesticize the means and personalize them in such a way that their emotions can flow toward the other person.

Six of our interviewees reflected explicitly upon the fact that the expression of emotions (and desires) through new technologies forced them to make their forms of communication more explicit. The possibility of playing with meanings, associations, sighs, and looks became less available since they had produced experiences with misunderstandings which, via the Internet or mobile, became much more difficult to solve. Two women argued that sometimes they had felt forced to expose themselves to their partners, in a sexual and an emotional sense, in order to overcome their partners' reproach of them having turned cold and unwilling to communicate emotionally about their love.

> There are so many things you don't have to say when you are face-to-face.... Over email, texting ... or even Skype ... you have to be so careful ... so explicit!!! (....) You can argue for a whole night because of a misunderstood facial expression, a silly joke.... I always try to express everything I feel as explicitly as I can ... and when I don't feel like saying I love you, I say it all the same. (Sandra, 38)

On Desires and Expectations

Desires and expectations regulate to a great extent the mediated communication between lovers, and help to generate positive and also negative emotions, according to their satisfaction or failure in attaining this satisfaction.

Our interviewees reported the ways in which, at the beginning of their relationships, for example, electronic communication had been an essential means of seduction. Long emails were sent back and forth, as well as brief love and desire messages:

> A simple 'I love you', or 'I wish you were lying naked next to me right now', you know? (Sandra, 29)

The first text and the first email, might have come as a surprise. The next were expected and desired. They were a challenge and, on some occasions, they became an obligation. The fine, fragile, and subtle line that separates surprise, challenge, and obligation became very clear in the interviews. Only mechanisms of self-regulation were of some help. Frustration, anger, and sadness often accompanied the initial feelings of excitement and joy.

> At the beginning we used to do it every day. Then we realised that it would destroy us, it was obsessing. We were forgetting our lives, our

jobs, everything. . . . And we decided to restrict it to once or twice a week. (Eva, 50)

The young man who had told us that mobile texts were like love pills also stated immediately afterwards that they sent five or six messages a day, no more. They did not want it to become inflationary. Sandra, to the contrary, became caught up in a spiral of sending messages expressing her longing and desire. She had started by asking the boy with whom she was having an affair for his reasons why he didn't want their relationship to become a stable love relationship, and she confessed:

> It is so easy to write what you think, what you desire, how hurt you are. You've got this pressure in your chest, you can't think of anything else . . . you just write and press the button . . . you know . . . it's just pressing the button . . . and then it's gone, it's sent. For a moment you feel better, relieved. Afterwards you think 'shit', and ten minutes later the anxiety of not having received a reply yet grows stronger than your own will. (. . .) You say to yourself every time, stop it; but it is so easy, and you know he will get it. . . . I sent thirty, perhaps forty mobile texts every evening. I begged, I insulted him, I called him a coward, I asked him to forgive me, I asked him why he didn't want to be with me, I told him that I loved him. He didn't reply. I felt like shit. (Sandra, 29)

Those who had used emails and texts in order to seduce one another at the beginning of their relationship, but had later slowly replaced these kinds of messages with organizational ones, told us how, at times, they looked back, and missed those messages, or the desire to write them, and used this change as a paradigmatic example for the changes their relationship had gone through over the years. The memories of the fantasies associated with the person hidden behind the email colored their present narration with nostalgia when the comparison was made between the possible futures that were dreamed of and imagined when writing and reading those emails and texts, and the reality behind the routines of quotidian family life, which was their present.

After analyzing the interviews, we would go so far as to conclude that electronic communication between lovers is strongly channelled, shaped, and influenced by a dangerous game with expectations and desires, which can be (and are) very quickly let down and disappointed. We have also observed that seduction is possible via electronic communication, highly effective, even. This is because it generates and transmits desire, but delays fulfilment of the same, at the same time as promising how wonderful this fulfillment will be when it takes place. However, we did not find any successful cases of a consolidated and stable relationship based on cohabitation, which, for whatever circumstances, had been 'forced' to become a long-distance relationship. In the cases we gathered, our interviewees only had stories of disappointment, disaffection, and growing apart to tell. They were not able to create the world of fantasy and seduction, the reign of momentary liminality, which couples who are just beginning a relationship seem to be able to create.

Couples who start their relationship online have to web and build their relationship from the very beginning using online rituals. They usually play with their partners and their own desires using masquerades and words regarding a possible joint future, or future moments when they might or will meet up. Here, electronic communication becomes a means for everything—for hot and cold moments, for the daily necessities, routines, and hot rituals. The emotions shaped in these rituals are from the very beginning adapted to the medium.

In face-to-face relationships this is different. Lovers have their hot moments when they meet physically, and even when they have some online rituals, these are not a necessity (or are not perceived as a necessity) for their relationship. The pressure on a certain moment of electronic communication is less high by far, the expression of emotions and the shaping of emotions in these few hot moments are not contradicting the emotional worlds webbed by the couple in their face-to-face daily life in such a strong way that the relationship is at risk of breaking apart. Communication via email and mobile phone can be used for organizational matters, and that is what it is used for in most cases, without calling the enchanting nature of the relationship into question.

In relationships which, after years of cohabitation, change into distance relationships, either the couples use electronic communication in the same way they always did, and thus miss out on those crucial moments of liminality that hold their relationship together, or they try to reproduce their former face-to-face rituals and routines online. This causes problems, as the bonds that can be (successfully) webbed through electronic communication and the emotional worlds that are implied in webbing bonds of love online differ from those that are webbed when two people are standing in front of each other. The change from face-to-face to distance (online) relations demands a transformation of the rituals, and therefore of the desires, expectations, and emotional worlds, a change that most couples find difficult to manage.

On the Authenticity of the E-Relation: General Assessments of Electronic Communication Within Couples

It is especially interesting to realize that we did not find a single person in our sample who asserted that he or she preferred electronic communication over face-to-face. All our interviewees, including those whose relationship would hardly be possible to maintain without electronic communication, presented electronic communication as being a hindrance to their relationship.

No, it certainly does not make things easier. To the contrary, it complicates it all, it makes it more difficult. (Mark, 29—the same person who had described Skype as the "love provider" of his relationship)

At this point we would like to add that this general dislike contrasts with some of the concrete assertions that we found in the interviews, when the interviewees were not explicitly replying to the question of whether these electronic means of communication helped in the relationship. Thus, for instance, we found statements such as:

When we are really, really angry, he always leaves the house. He goes away and after a while he or I start texting each other. At the beginning we're very angry, and just repeat the last insults we said to each other face-to-face. Then, it all cools down a bit, and we start discussing what has happened in a way that we have not done while being together . . . when we've talked things over, when it is all a bit better, he comes back and we end the discussion face-to-face. (Núria, 36)

Or:

Well . . . things I would not have said to her face-to-face . . . well I did tell her by email that I thought our sexual life was not satisfying . . . well . . . that we didn't have enough sex. I know it is wrong, (. . .), and she also told me that she had wished I would have told her face-to-face . . . but I couldn't, you know? Baby . . . you're not giving me enough, you know? I couldn't do that (. . .) by email . . . at least we had the topic on the table . . . and we could start talking about it. (Pere, 35)

In this last quote the "I know it is wrong" is especially significant. This is a point we found in our interviews unanimously: there are things you cannot (or better, should not) say via electronic communication. There was unanimous agreement upon the fact that one should not break up a relationship if it is not face-to-face if there is the possibility to do so (otherwise it has to be Skype to, "at least, look in the person's eyes when you say it"; Serena, 31)

Particularly the youngest interviewees affirmed that there are many things they would never have dared to say if it were not for the possibility of hiding one's face behind the medium. However, they also reported the negative effects of this 'need' to hide.

As well, all our interviewees referred to the "life together" that they could have with their partners when they were apart over the Internet as a simulated life.

We can even have some sort of sex . . . you know. . . . But it is not real. . . . It's just a simulation, a game with what you would like to do. It keeps the desire awake until you meet up again. (. . .) when we meet up again . . . we do not talk much. We just touch, smell, feel . . . all that which was not possible while being apart. . . . The relationship feels complete. (Ian, 38)

CONCLUSIONS

The analyses of the collected interviews have shown that love relationships nowadays are also being webbed through the means of electronic communication. And, even if the necessary ingredients for webbing a bond of love may remain the same, their webbing does occur in new places compared to the days before the generalization of electronic communication. Thus, love bonds may be webbed at work, on the train, two meters or two thousand miles away from their beloved. Love bonds can be webbed almost everywhere, as long as we have Internet coverage, and independently of whether we are or we are not physically next to our partners. This possibility of continuously webbing the love relationship, of continuously being in touch, sets the basis for another order of time and a different experience of temporality: you can be in touch 24 hours a day; you may sometimes engage in very momentary interactions that take place within a highly limited time frame, and compete with other forms of social interaction which evolve at the same time.

This transformation is not just a formal one, but a transformation that has reshaped and transformed the meaning of reciprocal actions and effects (*Wechselwirkung*) between couples engaged in the webbing of their daily bond of love, and it has transformed the meaning of those ingredients necessary for webbing bonds of love. These substantial changes lead to a change in the expression, performance, and experience of emotions and desires, as well as to a change in the expectations and needs of people in love.

These changes in the daily webbing of love relationships do not only affect long-distance relationships, which mostly depend on electronic communication for a fluid and frequent interrelation. Also, lovers who live together or close by have become increasingly used to electronic communication, without realizing it much, incorporating these immediate forms of relating to the other—in spite of the distance—into their taken-for-granted world. In these latter cases, electronic communication is mostly used for the quotidian organization of daily tasks, and for sporadic re-enchantment messages or, in some cases, for clarifying or romantic emails. In contrast, in long-distance relationships electronic communication is also used for creating those special moments, which extend their effects on the relationship much further beyond their temporal limitation, giving meaning and consistency to the whole of the relationship (finite universes of meaning).

Our analyses have also shown that the different devices, software, and applications used by lovers in order to communicate with their partners allow very different forms of practices, and very different forms of expressing and performing emotions, sensations, thoughts, and necessities, and therefore of webbing relationships. Thus, our interviewees often had a clear order and hierarchy of the different forms of electronic communication that were available to them. However, these hierarchies and specialization of uses were not the same within each couple. Specialization of devices and communication forms in relation to the practices and forms of interaction were a constant in all our interviews; the uses, meanings, and preferences varied. Thus, while Skype appeared as a means for performing rituals of love and seductive practices, and mobile messages and emails turned out to be used mainly for the organization of daily tasks for some couples, others referred to brief mobile texts as being the most seductive form of electronic communication, while others argued that nothing could substitute a long love email.

However, the specialization of the means of communication within each couple in order to cover certain communicational needs or desires was not the only constant element that could be identified in the interviews. The acceleration of the tempo of communication, and the acceleration of the emotional changes (happiness, disappointment; joy, anxiety; hope, despair) were also constants.

Furthermore, we realized that electronic communication was of vital importance in all the couples that we gained an insight into. For some, as a way of organizing daily life, for others as a means of sporadic re-enchantment of the relationship, while for others it was the very possibility of being regularly in touch with their beloved. Those people who lived with their partners and tended to use electronic communication for organizational matters seldom reflected on the use they made of electronic communication, and the vital importance that this form of communication had for their daily life. They had incorporated it into their taken-for-granted world. In those cases in which electronic communication meant the very possibility of having a relationship, the interviewees were very much aware of this fact, and despite acknowledging that they 'needed' electronic communication in order to relate to their loved ones, they did not have a high opinion of the kind of relationship it enabled. The preference for face-to-face relationships was clear, despite acknowledgments that some great and significant moments had been spent online.

New communication technologies have brought new possibilities and rules for webbing bonds of love to couples, whether they are engaged in long-distance relationships or living together. These possibilities have created new needs, desires, and expectations, and have given rein to their fantasies and imagination. However, our interviewees all agreed that no form of electronic communication could substitute a kiss and words of love breathed slowly into the ear of the beloved.

NOTES

1. Bauman 2003; Ben-Ze'ev 2004; Whitty 2003; Wildermuth and Vogl-Bauer 2007; Illouz 1997; Illouz 2007; Illouz and Finkelman 2009; Stafford 2004.
2. Compare this argument also with Friedland and Boden 1995.
3. On the subject of emotions and romanticism in *The Philosophy of Money* see also Frisby 1994: 139.
4. We had deliberately left the question regarding their uses of electronic communication for the end of the interview.

REFERENCES

Bauman, Z. (2003) *Liquid Love: On the Frailty of Human Bonds*. Cambridge, UK. Polity.
Ben-Ze'ev, A. (2004) *Love Online: Emotions on the Internet*. New York. Cambridge University Press.
Durkheim, E. (1995) *The Elementary Forms of Religious Life*. New York. Free Press.
Friedland, R. and Boden, D. (1995). *NowHere: Space, Time, and Modernity*. Berkeley. University of California Press.
Frisby, D. (1994) *Georg Simmel: Critical Assessments*. London et al. Routledge.
Illouz, E. (1997) *Consuming the Romantic Utopia: Love and the Cultural Contradictions of Capitalism*. Berkeley. University of California Press.
Illouz, E. (2007) *Cold Intimacies: The Making of Emotional Capitalism*. Cambridge, UK. Polity.
Illouz, E. and Finkelman, S. (2009) An Odd and Inseparable Couple: Emotion and Rationality in Partner Selection. *Theory and Society*, 38, 401–422. Retrieved September 30, 2011 from: http://www.jstor.org/stable/pdfplus/40345661.pdf?acceptTC=true.
Simmel, G. (1989) *The Philosophy of Money*. London et al. Routledge.
Simmel, G. (1999) *Gesamtausgabe in 24 Bänden: Band 16: Der Krieg und die geistigen Entscheidungen. Grundfragen der Soziologie. Vom Wesen des historischen Verstehens. . . . BD 16*. Frankfurt am Main. Suhrkamp Verlag.
Stafford, L. (2004) *Maintaining Long-Distance and Cross-Residential Relationships*. London et al. Routledge.
Strauss, A.C. and Corbin, J.M. (1998) *Basics of Qualitative Research: Second Edition: Techniques and Procedures for Developing Grounded Theory* (2nd ed.). Thousand Oaks et al. Sage Publications.
Turner, V. (1979) Frame, Flow and Reflection: Ritual and Drama as Public Liminality. *Japanese Journal of Religious Studies*, 6, 465–499. Retrieved from: http://www.jstor.org/discover/10.2307/30233219?uid=32051&uid=3737952&uid=2&uid=3&uid=67&uid=5909704&uid=32050&uid=62&sid=211025108 73811.
Whitty, M.T. (2003) Cyber-Flirting: Playing at Love on the Internet. *Theory & Psychology*, 13, 339–357. Retrieved September 21, 2011 from: http://tap.sagepub.com/content/13/3/339.full.pdf+html.
Wildermuth, S.M. and Vogl-Bauer, S. (2007) We Met on the Net: Exploring the Perceptions of Online Romantic Relationship Participants. *Southern Communication Journal*, 72, 211–227. Retrieved September 30, 2011 from: http://www.tandfonline.com/doi/abs/10.1080/10417940701484167#.UfpBfGQd5qI

Part IV
Mediating Emotional Space

10 Emerging Resentment in Social Media
Job Insecurity and Plots of Emotions in the New Virtual Environments

Elisabetta Risi

INTRODUCTION

This chapter synthesizes a few conceptual cruxes emerging from a theoretical and empirical study on the emotional dynamics that are developed and "vented" through online communication. The research considered the online social sharing of resentment through (typically narrative) user-generated contents (UGCs). Among the various emotions and feelings shared across the Web, I focused on resentment, as a heuristic category suitable for interpreting the various aspects of some present-day social relations and of the so-called *sad passions* of today's society (Benasayag and Schmit 2005).

The emergence of communication forms through digital media, as well as the close link between emotion and communication, have led to questioning the outcomes of the emotional language and content of the so-called Web 2.0.[1] Individuals who regularly use the Internet are increasingly turning into "authors" of what makes up the greatest part of the content available online. In this chapter the web has not been considered as a network of information and thoughts with no connection to emotions: many people's daily routines now include sharing their personal opinions online, writing comments in blogs, and posting messages in social networking sites. Such contents are seldom neutral—in fact, they are most often emotionally charged.

Online posts are normally short, condensed in a few words or made up of brief stories shared with other people in cyberspace. Biographical and narrative approaches (see Somers 1994; Bruner 2003) have shown that identities are shaped through the accounts of our lives found in the stories we tell, or that others tell about us. The intertwining of one's life stories and those of others creates a sort of biopic which contributes to making sense of one's own experiences and feelings.

Among the various sentiments, resentment could be defined as an emotional construct experienced by people in situations of injustice, and emerging in particular in present-day systemic contexts (Tomelleri 2010). It is regarded as the (frustrated) desire to reach an "object of value" (Girard 1999) such as a specific status, or certain resources or goals. Though these objects are potentially accessible in society or in one's *comparative reference group*,[2] a number of social actors are unable to obtain them owing to an entity (individual or collective, and more or less determined), which is hereby considered as the (real or perceived) culprit of the unfair situation.

The structural changes in the work system have brought about the progressive individualization of employment biographies and called into question the professional sphere, which had generally represented a key reference point for identity construction in modernity, as a means to establish mutual recognition and strengthen social ties. This life sphere is currently undergoing a necessary and radical redefinition (Sennett 1999), and, according to Beck (1992), social actors are forced to seek "biographical solutions to meet the systemic contradictions," in a context of widespread competition.

The research focused in particular on the so-called precarious workers, increasingly seen as typical of the current Italian work system.

The importance of the social sharing of emotions is confirmed by empirical research (Rimè 2009) and also observable when analyzing everyday conversations often reporting emotional episodes. The present chapter investigates the added value of sharing emotions and feelings on the Web by considering some "emotional" elements voiced and shared by temporary workers through UGCs. The emotions narrated are identified either explicitly (through verbal clues) or implicitly (by interpreting the elements of the stories and the mood of the online messages) and are assumed to play a role in the processing and evaporation of resentment.

After discussing the most important theoretical and conceptual aspects of the issue, the chapter presents a number of results of the empirical research.

RESENTMENT AS A CONCEPTUAL CATEGORY TO INTERPRET EXPERIENCES OF JOB INSECURITY

Late modern society, as Scheler had already argued (1975) for modernity, is egalitarian in terms of the declared rights and values, but it is characterized by a new inequality as to the possession of tangible and intangible resources. Despite promises of a bright future and never-ending progress, *late modernity* (Giddens 1991) is still marked by structural conditions of social and economic inequality. Moreover, the current economic crisis and the contradictions generated by the labor market over the decades have exacerbated experiences pertaining to "new" areas of social weakness, among which is job insecurity.

Work—and in particular the social division of labor—was the heart of classic sociological analysis. Writers such as Marx and Durkheim dealt with this phenomenon, albeit from different perspectives, as a major element in social organization and the construction of modern-day identity, mainly associated with one's status or social class of birth (e.g., Mingione and Pugliese 2002).

Late modernity has abandoned the traditional hierarchical model of society (Sennett 2006) which for years had accompanied the life experiences of workers. The move away from a Fordist model has resulted in recent years in a gradual transformation of occupational conditions with various outcomes, such as processes of relocation, lack of efficient mechanisms for providing new jobs to the unemployed, the deregulation of the labor market, and widespread more or less institutionalized competition.

Today's society has seen a growing diversity of job types and also the rise of new kinds of economic actors at the intersection of a new capitalism and new technologies—all this in a social context of new post-Fordist working arrangements, characterized by more privatized, flexible, and precarious occupational models (Fisher 2010).

The gap between growing expectations of achievement and subjective professional experiences has been widened in a context of new emerging social risks. The idea of *social risk* is connected to the need to make continuous career and life choices in the scenario of *reflexive modernization* (Giddens 1991; Beck 1992), in which the elements of social life are constantly brought into question. The (untranslatable) conceptual category of *Unsicherheit* can be used to describe this scenario, which combines in itself the sense of insecurity and instability due to the absence of social guarantees, as has been widely theorized and debated (Bauman 2000; Beck 1992). Among the different *risk profiles* (Ranci 2010) of contemporary society, this research focused on the predicament of precarious workers. This notion refers to a very complex set of occupations and activities that, though different, share the same "employment flexibility (...). A wide assortment of job contracts, called atypical to distinguish them from the typical permanent and full-time ones" (Gallino 2007: 6).

Job insecurity itself could be considered as a "precarious" concept, since its object is flexible and loosely defined: indeed it encompasses a plurality of people working under short- or medium-term, non-standard contracts, and others seeking work or having experienced fixed-term periods as occasional workers (Pedaci 2010; Borghi et al. 2001).

This study analyzed resentment because it seems the emotional construct most related to the experience of injustice as perceived by some social groups, namely that of precarious workers.

The more philosophical speculations on resentment, in the meaning of *ressentiment*, were conceptualized by Nietzsche (1887) and Scheler (1975). Their theories are worth mentioning—even though to distance ourselves from them—because they had the merit of identifying the grievance with a theoretical concept that can be used to describe socio-cultural processes. In spite of their differences and divergences (Meltzer and Musolf 2002), those philosophers assigned to *ressentiment* a crucial role in the creation of modernity and in the historical transformations between the 19th and 20th centuries.

Among the various affective conditions, resentment is one that scholars from different fields have recently been interested in (among others,

Barbalet 2002; Girard 1999; Angenot 2010), one possible reason for this being the wide diffusion of this feeling in present-day society.

From a constructionist point of view, values, needs, and desires are considered (except for the survival-related ones) as historically- and culturally-specific social constructs, built and shared within a society. Whenever mimetic processes are triggered off among individuals—e.g., the desire to be (like) the other (Girard 1999)—the social actors seem to experience some negative emotional charges, in particular one of resentment (Tomelleri 2010).

Emotions are amorphous categories, which tend to merge into each other and appear as complex structures, including more than one feeling at the same time, in succession or in very close temporal proximity (Barbalet 1992; Benski 2005, 2011; Collins 1990; Scheff 1979), and with a tendency to form into constellations. Complex situations give rise to complex emotional experiences, which quite often cannot be adequately accounted for by focusing on a single emotion, so speaking in terms of constellations of emotions (Benski 2011) is the most appropriate way of exploring this phenomenon.

In the English language the notion of resentment has become synonymous with anger, spite, and bitterness. According to Solomon (1993) resentment could be assigned to the same conceptual continuum as contempt and anger: while resentment is directed toward higher-status individuals, anger is directed toward equal-status ones and contempt toward lower-status ones.

The Italian word encompasses different feelings and can express several shades of emotions: it is therefore important to consider the constellation of emotions that gravitate around resentment. This constellation can take on various forms and be composed both of congruent emotions (which amplify the predictable behavioral consequences), and a mixture of contradictory or non-congruent emotions (e.g., anger and shame), in which again the behavioral consequences are different from the emotion rule expectation of each of these emotions alone (Benski 2011; Yang 2000).

This theoretical framework assumes emotions to be generated and (re)produced in daily social contexts and their meaning to be culturally and socially built. They also tend to form into constellations that take various forms and can have different consequences.

Moreover many emotions and feelings have a relational nature: resentment, in fact, has its generative source in the mimetic comparison between individuals and in a relational element which people wish to take it out on, since it has caused damage that they cannot counteract.

From a macro-sociological perspective, the economic and social structures in advanced Western societies could be potential generators of experiences of resentment. This feeling could, in fact, represent the emotional response specifically positioned between the (micro) individual experiences and the (macro) contradictions and dynamics of the social system (Beck

1992). In particular, the notion of resentment has proved useful to interpret those situations related to the perception of being the victims of some sort of social injustice, in globalized and highly flexible contexts.

Over the centuries, cultures have developed symbolic devices, practices, rituals, and institutions to voice, interpret, and channel the negative energy of social resentment. Political analysis, and social conflict in particular, seem to be the rational frameworks that gave voice to the development of resentment in modernity. Class conflict has, in fact, been considered as a kind of (social and) political organization of feelings of anger and resentment, connecting them to the sense of justice and social equality (Barbalet 1992; 2002).

If the framework of modernity shows a hierarchical view of power, according to Marshall's analysis of class resentment (1973) the diffusion of formal democracy and egalitarianism in the management of power should mitigate resentment; nevertheless this mood does not disappear (Barbalet 1992) and contemporary society does not therefore seem immune from the spread of social resentment.

According to some hypotheses (Tomelleri 2010), the systems of social protection and the patriotism of modern national states were *compensation devices* for the containment and sublimation of social resentment. At present, however, the European welfare systems, secured in the states of the continent during the last century, have seen a remarkable reduction in the mechanism of social protection, and the current economic development in a globalized context has increased social differences and inequalities. In this framework, therefore, the previously-mentioned compensation mechanisms no longer seem to be effective in our individualistic society, so that the elaboration of social resentment rather appears to be part of the lives of people situated in specific contexts and social relations.

Resentment occurs in particular in the structural contradictions typical of capitalist and democratic societies, where certain situations are considered as "socially unfair," despite a cultural background that proclaims the equality of all individual desires. Moreover, the process of individualization has led to a growing sense of responsibility as to the construction of people's biographies and careers: social actors are required to be mobile, flexible, not bound by membership, using a "dynamic model of action" (Beck 1992) in order to be able to access a number of hypothetical job opportunities. However this process appears to the subjects more as a source of anxiety than as actual freedom of choice.

Hyper-individualism (Lipovetsky 2006) takes to extremes the process of emancipation of individuals from the bounds of membership (class, traditional, religious), thus giving such individuals total responsibility for their own lives and hence for their own possible failure. Under pressure from a highly competitive context and a cultural-symbolic system that reveres the total satisfaction of desire, involved in mechanisms that are rarely rewarding on the basis of their merit and personal abilities (Sennett 2006), social actors end up experiencing a constellation of negative emotions connected to the unfair situations that they feel they are suffering.

The various aspects discussed so far, considered as potential sources of resentment, make up the background of the research, which refers to the recent sociological theories on contemporary Western society. In the next section a number of Internet environments and UGCs will be interpreted as symbolic and social devices through which resentment was expressed, shared, and elaborated.

THE ONLINE SOCIAL SHARING OF EMOTIONS

Specialized literature—especially in the psycho-social field—has highlighted the negative emotional components that occur in some circumstances. According to Rimè (2009), human beings are continuously engaged in goal-reaching activities, but if the pursuit of a goal is substantially slowed down, or if it is blocked, this results in a negative emotional state (Carver and Scheier 2001).

In such cases the value of the social sharing of emotions is not only linked to its role for mitigating personal suffering and the frustration deriving from negative emotional experiences: sharing emotional experiences also becomes a way to access a socially constructed reality and its effects can therefore go beyond the individual level.

Numerous data confirm that experiencing negative emotions leads to individual cognitive work on the one hand, and on the other to social sharing—the narration and conversation with others. Bruner (1991, 2003) sees narration as playing a central role in the context of emotional circumstances: the construction of a story is useful to make sense of the experience, creating a plot that makes events understandable (to oneself and others). By narrating events and sharing them with others, human beings construct a collective memory and create networks of sharing that bear the mark of social experiences (Rimè 2009).

The social sharing of emotions often occurs through the presence of an interlocutor, even though the latter may be there only virtually. The need to share one's feelings is not related to the illusion of freedom from suffering or frustration, but rather to finding relief in exchanging views with others. In addition, sharing emotions can be a tool for stimulating bonding and strengthening social ties (Rimè 2009).

The social practices connected with Web 2.0 have shown that quite often it is not information and data that are "exchanged" online, but rather relational content, fragments of life stories and consequently emotions, feelings, moods.[3] Therefore the Internet—and in particular certain online interactive environments—could be conceptualized as a socio-technological artifact capable of connecting individuals and generating not only a *collective intelligence* (De Kerchove 2001), but also an emotional network between web users. The Internet has been described as a repository capable of collecting, capturing, and directing some emotions, whose use in daily life contributes to the emotional life of individuals and to the restructuring of their emotional repertoires (Vincent and Fortunati 2009). But beyond its representation as a virtual space where emotions and feelings are simply deposited, the web is above all a place for sharing such emotions and feelings, which would hardly be possible otherwise owing to physical or social distances.

More and more people are now able to create and share contents over the web and research has shown that social networking sites attract an increasing number of users (Ito 2008). For a few years now social sciences have therefore had to take account of all this narrative material (both in multimedia and, more often, in narrative form), which becomes *permanent, searchable, replicable* (boyd & Ellison 2007), and made public by net users.

Further studies of the processes by which a socio-technical structure allows emotional states to be rendered objective, elaborated, and shared (a sort of *web emotionology*), could provide better understanding of the recent success gained by some social networking websites (Facebook, Twitter, and the like).

The empirical research described in the present chapter was carried out by reading and analyzing the content of online posts: the vast majority of them are short or long stories in blogs, or messages published in forums and social networking sites. The narrative UGCs that were taken into account objectify the emotional experiences of net users and become "tracks" of their lives, stored and shared online with others.

The study investigated the value of the online social sharing of emotions, namely the relational processes of expressing and sharing resentment (individually and collectively felt in situations of job insecurity) and the role of UGCs in the mechanisms of elaboration and possible dissipation of this feeling.

Given that one of the outcomes of the social sharing of emotions is the strengthening of social ties between social actors (Rimè 2009), we also tried to investigate whether and how this might be actualized in the so-called *weak ties*[4] (Granowetter 1981) built through the net.

THE EMPIRICAL RESEARCH: PATTERNS OF RESENTMENT ON THE WEB

The empirical analysis focused on the contents and forms of a sample of communicative and narrative texts, spontaneously generated, published, and shared online by users in a number of online environments created around the issue of precariousness and job insecurity. Such online spaces appear to be used by precarious workers to 'deposit' contents about their experiences and emotions, occasionally even in real time.

The empirical research was conducted through a qualitative analysis of the contents of about 850 posts published on a sample of 10 blogs, 5 discussion forums, and 10 social networks (Facebook pages and groups).[5]

Through a narrative type of approach (Bruner 1991; Riessman 2002), the interpretative analysis of the texts reconstructed the main contents, the textual forms used and the narrative contexts within which the emotions were embedded.

The results to be dealt with in this chapter provided the following indications:

1. More or less evident traces of resentment, in some cases explicitly expressed through verbal content, but mostly inferred from the narrative forms and elements of the stories told;
2. Some "prototypical" stories underlying the events narrated by the net users and their connection with specific elements that seem peculiar to experiences of resentment;
3. The role that some social media where emotions can be shared may play

   - in the elaboration, dissipation, or sublimation of social resentment;
   - in relational processes between social actors and in formulating actions of collective mobilization by precarious workers.

Looking for Resentment and Its Narrative Structure

In spite of the fact that sharing emotions and feelings online may be more uninhibited, resentment is seldom explicitly mentioned in web users' reports of their personal work and life experiences. Expressing negative emotions is already difficult in itself (Rimè 2009), but mentioning resentment seems to be even harder. On the one hand this may be due to the special complexity of this *unmentioned* feeling and the fact that people are less clearly aware of it (Kancyper 1991). On the other, the fact that the very word 'resentment' hardly ever appears in the posts considered in this research points to its composite nature, given that in order to describe the feeling one needs to resort to other words denoting other emotions. It is therefore all-important to focus attention on the expressions and narratives where such words occur so as to better understand the actual emotions and allusions.

In the stories of precarious workers the feeling of resentment is instead evoked through a number of emotions and moods associated with it. This is done in particular through different combinations of anger, bitterness, disappointment, shame, hate, and those other dysphoric emotions that arise from the sense of frustration and powerlessness in the face of an experience one has been through.

Since various combinations of the previously-mentioned emotions often occur in conjunction, constantly merging into each other (Barbalet 1992; Scheff 1979), some scholars group them together and refer to them as the "emotion family" of anger (Hochschild 1983; Lazarus and Lazarus 1994). According to Benski (2005, 2011), the notion of "emotional constellations" may prove useful when trying to understand the complex structure of some emotions and moods and their behavioral expression. On the basis of these insights, resentment may be considered the "lodestar" of a constellation of emotions that takes on different forms in the various environments analyzed. The feeling cannot be adequately accounted for by focusing on the single emotional state, but is rather more thoroughly explorable through the narrative forms in which it crops up.

The narrative dynamic and the emotional tonality surfacing in the UGCs considered in our analysis make for an interpretation of these contents as stories of resentment. These, in fact, trace out the path of "injustices suffered" in the working world, characterized by an asymmetrical relationship between the narrators—in most cases the protagonists of the story—and someone/something equipped with powers, who/which creates frustrating situations for these precarious workers and typically prevents them from reaching their goals and aspirations.

In the samples observed, resentment experiences had been shared through long or short narrative texts, this being the predominant form in online environments. Therefore some *prototypical* stories of precarious workers were identified by analyzing how resentment, albeit implicitly, emerges through some characteristic emotional elements, considered to be the structural conditions for the rise of this complex feeling. More particularly, the UGCs found in blogs, forums, and social network sites emphasize the following elements :

- The objects of value: the research pinpointed a corollary of objects of value, all apparently centered around the macro-notion of "security" as opposed to that of *Unsicherheit*.
- Perceived injustice: precarious workers describe their personal experiences as unfair in relation to their actual merits and to an ideal "social justice."

We 40-year-old unemployed men and women with excellent references and CVs cannot help feeling miserable: we are clearly victims of injustice. What can we do to find a job? I feel helpless—it's been like this for two years now.

I'm feeling angry, disappointed, resentful. . . . I've been mostly unemployed for the last few years. . . . Now with the economy in crisis I'm still precarious. My employment contract is project-based and expires in 6 months: that's totally unfair! I'm even considered as a mummy's boy as I can't afford to take out a mortgage, marry my girlfriend or . . . go anywhere for the next summer holidays!

- The victims: precarious workers describe themselves as victims. This notion is linked to the previously-mentioned perception of injustice, which seems to *justify* some negative or even aggressive feelings and behavior. The temporary job model may well be functional to the current logic of production economy, but it definitely makes people feel penalized in more than one way in their work and life experience. Though highly-qualified and experienced, these workers, mostly young adults, are obliged to accept humiliating working conditions—temporary contracts, relatively low salaries, heavy work schedules, etc.—unsuited to their level of education and offering no guarantee for the future.
- A plurality of offenders: it is hard to pinpoint a single actual culprit against whom to vent and channel one's resentment. The blame is in many cases laid on the "system"—be it political, economic, or institutional. The "system" is a narrative device which the UGCs under consideration have identified in particular with politicians (Berlusconi for one), the state, the enterprise or the owners of the company in question.
- Some scapegoats: as no one in particular can be blamed for causing their miserable situation, the victims transfer at least part of the blame, in their online narratives, onto a number of scapegoats, the most frequently quoted being multinational corporations, supranational bodies, and above all, migrant workers. Such symbolic elements seem to suggest the attempt to remove and channel their negative emotional energy against subjects seen as distant from their personal experience: "The immigrants are to blame! It's full of Egyptians here in Italy, along with their families, and they steal our jobs!"

An interesting point that the research indicates is the polarization of the "blame." While on the one hand many precarious workers accuse the institutions, politicians, or other scapegoats characterized in negative terms, on the other quite a few online stories and statements lay the blame on the precarious workers themselves, describing them as "uninterested, and above all incapable of organizing any common action." This aspect seems to undermine the possible creation of an internal group solidarity among precarious workers.

As they cannot identify any actual culprit for their unsatisfactory work situation, the protagonists of some online stories end up blaming themselves for it, even though the causes are clearly of social and structural origin: "Well . . . here the blame goes away, it widens and then narrows, then widens and narrows and eventually I myself become the one to blame."

In these plots, the precarious workers UGCs acquired also a bitter sense of disappointment and sometimes of resignation about social change.

Elaboration and Effects of the Online Sharing of Resentment

This study has endeavored to investigate the value and role of online communication and social relations in the processes of elaboration of resentment.

To begin with, the stories told by precarious workers in online social environments appear to have a communicative function related to:

- The expression of one's discomfort—often jotted down without much thinking—simply to vent one's feelings through a written narrative that may make them sound objective: "I doubt I'll manage to explain how I feel: I've had this anger and pain inside me for years and it's so hard to put it into words."
- The statement of one's negative experiences with a view to making them public and denouncing them: "I wish this space would turn into a cry loud enough to be heard by those who are ignoring us"; "I'm here to shout out our anger and denounce our condition."

At the root of these two functions lies the importance of the social sharing of emotions made possible by the generation and publication of contents and stories on the web. Sharing resentment online then becomes a practice that may in particular alleviate the bitter feelings of precarious workers. They share their distressing work experiences online in order to find some comfort in the online interaction with other net users.

The storytelling through writing seems to be more functional to venting one's feelings; it would be interesting to investigate this practice (with further study, through in-depth interviews with a sample of temporary workers) as an occasion for *deep acting* (Hochschild 1983). Deep acting concerns the changing of both internal experience and the public display of emotions: writing and sharing online UGCs could, in fact, be a technique to manipulate the components of emotion before it is fully underway and to evoke or retract the appropriate feelings and images at the appropriate time.

Several of the posts describe these forms of online communication as a possible escape route from contexts where face-to-face sharing of emotions is complicated and difficult. One reason for this difficulty is that experiences of insecurity are "dispersed," since precarious workers are themselves scattered in different work sectors and companies. And yet, even when the subjects share the same workplace—as is the case of precarious workers in one single call center—direct personal interaction is apparently discouraged by other factors such as competition among colleagues, especially in jobs where high staff turnover and flexibility are the rule.

On the web distance and competition seem to disappear: the implicit or explicit recipient of the messages and UGCs posted by precarious workers are mainly other temp workers, visiting these web environments as regular or occasional users, sometimes acquaintances but mostly strangers to each other. Joining in a web community or in other symbolic online networks can lead to processes of *recognition* based on the fact of sharing the same harsh work situation: "The Internet brings people together more than any other single group, company or union . . ."; "If the Web didn't exist we wouldn't be able to talk out our problems or compare notes with such a number of people. We realize there are lots of us facing the same problems."

The analysis seems to substantiate the fact that the web can foster relationships especially in the form of ego-centered networks (Castells 2000, 2001), namely networks of social ties capable of providing support, a sense of belonging and some kind of attempt to construct social groups and identities.

The data collected, however, show that the social sharing of emotions and resentment on the web gives rise to "weak ties" between subjects, as a potential group identity and one in search of shared characteristics. The stories told online by precarious workers can in any case be interpreted as weak in terms of symbolic identities, but strong in terms of their expectations and of the expression of perceived injustice.

Some of the posts suggested the need to make concrete proposals for change, in conjunction with the social action undertaken in order to sensitize public opinion on the subject of precarious work. Mobilization actions should be organized both online—e.g., by sending mass emails, inviting people to sign online petitions, or urging them to visit and "like" Facebook pages—and outside the web, by calling strikes and street protests.

Scholars have also identified the web as a crucial resource for activists,[6] who have been provided with a means of communication extending beyond mainstream media and capable of creating information spaces for social conflict through the so-called "cyberprotests" or other such forms of collective action (Van Laer and Van Aelst 2009). Seeing how important public discussion of work precariousness has become and since it is now possible to mobilize masses of people, the protest against job insecurity can be considered as an interesting attempt to build collective identity and to seek visibility (Della Porta 2008).

Furthermore, the *emotional energy* (Collins 1990), combined with the perception of social inequalities, has played a key role in the formation of social movements, e.g., Flam and King 2005), and of collective and class identities (Runciman 1972; Barbalet 2002, 1992).

Within this framework, the results of the research have nevertheless shown that the online sharing of resentment—as well as of emotions—seems unable to strengthen the previously-mentioned weak ties beyond the mere act of joining online environments. It appears that in the Italian social context the emotional constellations of precarious workers have only occasionally eased the way from online proposals to social projects or actions.[7]

The various mobilization actions are often "announced" and overexposed through *viralization* in the several social networking websites: strikes and proposals for merging different groups of temporary workers are suggested and advocated, but in the end few traces can be found of the actual organization and effects of the protest, or of attempts to create a social movement. Although *meeting* and *recognition* among precarious workers do occur in online environments, the question of actually building a collective identity remains open and only potential.

One reason for all this seems to be that the masses of precarious workers have difficulty—as was mentioned earlier—finding a clear target against which to mobilize. According to Formenti (2009), it is also the increasingly fuzzy idea of work that makes it hard to identify the worker's "enemy," as this can no longer be associated with a single, well-defined class and/or institution, but is embodied by a complex and layered network of operating powers.

On the other hand, however, the actual lighting of the emotional fuse may be hindered by the keen competition on the workplace and the fragmentation of the workforce. These particular workers no longer consider themselves capable of creating strong ties—even beyond online networks—and a true group identity.

A number of macro-social issues may also be the source of these dynamics. Among them are: (a) the decline of ideologies, which were once able to bring together people's individual stories into macro-narratives by means of a common language, thus turning single "cases" into a "cause"; and (b) the process of *disembedding* (Giddens 1991) and that pressure of individualism, whereby "while single individuals may well admit having some interests in common with others, collective action implies acknowledging their own personal inadequacy. In fact you would not have to resort to collective action if you yourself were really talented" (Sennett 2006: 406).

Considering the previous discussion on the role of UGCs in processing the emotional constellation related to resentment, it is quite possible to claim that the energy generated by such emotions is therefore weakened rather than amplified as might be expected. Resentment seems to appear in emotional constellations that neutralize some of the expected effects because it is accompanied by contradictory emotions: indeed temporary workers also mention the shame. It will be interesting to investigate the presence of this feeling (shame), as seems characteristic of an individualistic society that gives more responsibility to individuals, so that even they are ashamed of being victims of unfair situations (which have clear social roots).

The resentment associated with a precarious work situation does not disappear, but the negative emotional baggage seems to be sublimated through the online narration.

This aspect can be interpreted through the "theory of catharsis" (Scheff 1977, 1979; Turnaturi 1995): temporary workers write and share online their thoughts, emotions, and experiences, through a sort of *ritual* practice. So they move their focus from the source of resentment to their own "safe" position of observers/readers, who distance themselves from the online story/post. The online storytelling produces a "distance" between themselves and their feelings: UGCs tend to play a role that is mainly expressive and sublimating to the resentment in question.

CONCLUDING REMARKS

Nowadays the Internet, and in particular a few social networking websites on the condition of precarious employment, are environments where more and more people, and precarious workers among them, tend to pour out their stories and experiences, benefiting from the opportunity to "meet" others and to share their emotions. In the UGCs analyzed, resentment is often evoked by expressing the constellation of emotions that accompany it through prototypical stories (consisting of certain characteristic ingredients: value object, victim, offender, scapegoat).

While expressing negative feelings through the practice of storytelling is in itself beneficial in an individual cathartic process made possible by writing, the social sharing of the resentment of temporary workers may also be accompanied by the suggestion of several proposals for mobilization, protest, and social action in general. However such proposals seem to remain mostly just announced and *viralized*, and are only occasionally carried out in practice.

These dynamics end up resembling a sort of rumination—one of the processes that nurture the resentment mechanism—which consists in the continuous mental recall of one's experiences and involves repetitively focusing on one's sorry plight, to the point that the subjects are unable to take any direct action in order to assert their rights.

Going beyond the subjective nature of the psychological approach, we might consider this concept in a more collective sense, as a sort of *social rumination* made possible by the increasing number of online stories, which narrate and restate situations of injustice experienced by temporary workers and suggest various concrete initiatives to expose and protest against this social problem. The protagonists simply continue to fantasize about the actions that they could—one day—carry out in order to obtain justice.

Online social networks seem to be turning into "expressive communities" in which to describe, express, and share one's emotional experiences, and considered as environments where precarious workers can vent their frustration and find some comfort.

Even if the online social sharing of emotions enables one to elaborate on and dissipate resentment at the individual level—breaking free from the shackles of mere mental rumination—the construction of plots that result from sharing resentment on the web seems to produce a sort of *social rumination*: single items, although expressing the same instances, find it hard to merge into one macro-story and to generate a shared social project on or against precarious work.

In some cases citizens have been mobilized and have created the initial critical mass, above and beyond the intervention of the media or of traditional social structures (such as political parties). The resentful precarious workers who have gone beyond mere complaint or social rumination have generated online groups not only in the form of expressive communities, but also as a basis for social movements capable of making their voices heard and to engage in political action.

When sharing resentment from the micro level of the online relation does not confine itself to expressing and venting one's own feelings, it could become an instrument of political mobilization.

NOTES

1. A Web 2.0 site allows users to interact and collaborate with each other as creators of user-generated content, in contrast to websites where users are limited to the passive viewing of content that was created for them.
2. A group which individuals use as a standard or point of reference in making evaluations or comparisons of themselves and of other individuals or groups. In the area of reference group theory the concept of relative deprivation has been explored, by examining various contradictions and feelings of satisfaction or deprivation in groups (see Runciman 1972).
3. An interesting review of the psycho-social literature on emotion in computer-mediated communications is Derks et al. 2008.
4. According to Granovetter (1981) strong ties are family, friends, and other people you have strong bonds to, while weak ties are relationships that transcend local relationship boundaries both socially and geographically. The "strength" of weak ties lies in the flow of information (e.g., he shows that jobs are more often found through weak ties than through strong ties).
5. The research sample is a qualitative one: starting from the search engine results pages in response to a keyword query on temporary work, the online environments analyzed have emerged as the most frequently "visited" by temporary workers and rich in UGCs.
6. However some researchers have shown that although participants in a transnational movement closely identify with, may even feel passionate about and/or gain an exhilarating feeling from membership, the movement remains dominated by the role of local activists (Della Porta 2008).
7. Some relevant exceptions are the episodic manifestations and rallies of some social movements that have used new forms of present-day protest in a context where the traditional "intermediate bodies" (political parties and trade unions), are progressively losing weight. An Italian example of such movements are: "The Purple People," which originated from a blog (http://www.ilpopoloviola.it/) and became visible in the public squares during the event of "No-Berlusconi Day" on December 5, 2009; "Our Time is Now" (http://www.ilnostrotempoeadesso.it) focused specifically on job insecurity. In these instances, the development of resentment has a generative perspective, which moves on from the online proposals to some concrete projects, often on a local basis.

REFERENCES

Angenot, M. (2010) Resentment. *AmeriQuests*, 103(1), 1–39.
Barbalet, J.M. (1992) A Macro Sociology of Emotion. Class Resentment. *Sociological Theory*, 10, 150–163.
Barbalet, J.M. (2002) *Emotion, Social Theory and Social Structure. A Macrosociological Approach*. Cambridge: Cambridge University Press.

Bauman, Z. (2000) *Liquid Modernity.* Cambridge: Polity.
Beck, U. (1992) *Risk Society: Towards a New Modernity.* London: Sage.
Benasayag, M. and Schmit, G. (2005) *The Age of the Sad Passions.* Milan: Feltrinelli.
Benski, T. (2005) Breaching Events and the Emotional Reactions of the Public: The Case of Women in Black in Israel. In D. King and H. Flam (Eds.), *Emotion in Social Movements.* London: Routledge, (pp. 57–78).
Benski, T. (2011) Emotion Maps of Participation in Protest. *Research in Social Movements, Conflicts and Change,* 31, 3—34.
Borghi, V., Chicchi, F. and La Rosa, M. (2001) Empirical Analysis of the Risk of Social Exclusion of Long-Term Unemployed Young People in Italy. In T. Kieselbach (Ed.), *Living on the Edge. An Empirical Analysis on Long-Term Youth Unemployment and Social Exclusion in Europe.* Opladend: Leske e Budrich, (pp. 319–391).
boyd, d & Ellison, N. (2007) Social Network Sites: Definition, History, and Scholarship. *Journal of Computer-Mediated Communication,* 13 (1), 210–230.
Bruner, J. (1991) The Narrative Construction of Reality. *Critical Inquiry,* 18, 1–21.
Bruner, J. (2003) *Making Stories: Law, Literature, Life.* Harvard University Press.
Carver, C.S. and Scheier, M.F. (2001) Optimism, Pessimism, and Self-Regulation. In E.C. Chang (Ed.), *Optimism and Pessimism: Implications For Theory, Research, and Practice* (pp. 31–51).Washington, DC: American Psychological Association.
Castells, M. (2000) *The Rise of the Network Society, The Information Age: Economy, Society and Culture* (vol. I). Cambridge, MA; Oxford, UK: Blackwell.
Castells, M. (2001) *The Internet Galaxy, Reflections on the Internet, Business and Society.* Oxford: Oxford University Press.
Collins, R. (1990) Stratification, Emotional Energy, and the Transient Emotions. In T.D. Kemper (Ed.), *Research Agendas in the Sociology of Emotions* (pp. 27–57). Albany, NY: State University of New York Press.
De Kerkchove, D. (2001) *The Architecture of Intelligence.* Basel and Boston: Birkhäuser.
Della Porta D. (2008), Protest on unemployment: forms and opportunities, Mobilization: an International Journal, 13(3): 277–295.Derks, D., Fischer, A.H. and Bos, A. (2008) The Role of Emotion in Computer Mediated-Communication: A Review. *Journal of Computer in Human Behavior,* 24(3), 766–785. doi:10.1016/j.chb.2007.04.004
Fisher, E. (2010) Contemporary Technology Discourse and the Legitimation of Capitalism. *European Journal of Social Theory,* 13, 229–252. doi:10.1177/1368431010362289
Flam, H. and King, D. (Eds.). (2005) *Emotions and Social Movements.* London: Routledge.
Formenti, C. (2009) Cognitive Capitalism, Crisis, and Class Struggle. The Post-Operaista Paradigm. *Sociology of Work,* 115, 131–143. doi: 10.3280/SL2009-115007
Gallino, L. (2007) *The Work is Not a Commodity. Against the Flexibility* [Il lavoro non è una merce. Contro la flessibilità]. Roma: Laterza.
Giddens, A. (1991) *The Consequences of Modernity.* Stanford University Press.
Girard, R. (1999) *Resentment. The Failure of Desire in Contemporary* [Il risentimento. Lo scacco del desiderio nell'uomo contemporaneo]. Milan: Raffaello Cortina.
Granovetter, M. (1981) *The Strength of Weak Ties: A Network Theory Revisited.* Albany, NY: State University of New York, Department of Sociology.
Hochschild, A.R. (1983) *The Managed Heart—Commercialization of Human Feeling.* Los Angeles: University California Press.

Ito, M. (2008). Introduction. In K. Varnelis (Ed.), Networked Publics (pp. 1–14). Cambridge: MIT Press.Kancyper, L. (1991) Remorse and Resentment in the Sibling Complex. *Revista de Psicoanalisis*, 48, 120–135. doi: 0034–8740

Lazarus, R.S., & Lazarus, B.N. (1994) *Passion and reasons: Making sense of our emotions*. New York: Oxford University Press.

Lipovetsky, G. (2006) *A paradoxical happiness. Considerations on hyper-consumption society*. [Le bonheur paradoxal. Essai sur la société d'hyperconsommation], Paris: Collection NRF Essays, Gallimard.

Marshall, T.H. (1973) *Class, Citizenship, and Social Development : Essays*, Westport, CT: Greenwood Press.

Meltzer, B.N. and Musolf, G.R. (2002) Resentment and Ressentiment. Sociological Inquiry, 72, 240–255. doi: 10.1111/1475-682X.00015.

Mingione, E. and Pugliese, E. (2002) *Work*. Rome: Carocci.

Nietzsche, F. (1887/1996) *On the Genealogy of Morals* (D. Smith, Ed. and Trans.). Oxford: Oxford World's Classics.

Pedaci, M. (2010) The Flexibility Trap: Temporary Jobs and Precarity as a Disciplinary Mechanism. *WorkingUSA*, 13(2), 245–262.

Ranci C. (Ed.). (2010) *Social Vulnerability in Europe. The New Configuration of Social Risks*. Basingstoke: Palgrave MacMillan.

Riessman, C.K. (2002) Narrative Analysis. In A.M. Huberman and M.B. Miles (Eds.), *The Qualitative Researcher's Companion*, (pp. 217–270). Thousand Oaks, CA: Sage Publications.

Rimé, B. (2009) Emotion Elicits the Social Sharing of Emotion: Theory and Empirical Review. *Emotion Review*, 1, 60–85. doi: 10.1177/1754073908087189

Risi, E. (2011) Rumors: Online Interactions Between the Public Sphere and Third Places. *Italian Review of Sociology*, 1, 87–116. doi: 10.1423/34335

Runciman, W.G. (1972) *Inequality and Social Awareness. The Idea of Social Justice in the Working Classes*. London: Enaudi Editions.

Scheff, T.J. (1977) The Distancing of Emotion in Ritual. *Current Anthropology*, 18, 483–505. doi: 10.1086/201928

Scheff, T.J. (1979) *Catharsis in Healing. Ritual and Drama*. Berkeley, CA: University of California Press.

Scheler, M. (1975) *Ressentiment*. (L.A. Coser, Ed.; W.W. Holdheim, Trans.). New York: Schocken.

Sennett, R. (1999) *The Corrosion of Character, The Personal Consequences Of Work In the New Capitalism*. New York and London: Norton.

Sennett, R. (2006) *The Culture of New Capitalism*. New Haven, CT: Yale University Press.

Solomon, R.C. (1993) *The Passions: Emotions and the Meaning of Life*. Indianapolis, IN: Hackett Publishing.

Somers, M.R. (1994) The Narrative Constitution of Identity: A Relational and Network Approach. Theory and Society, 23, 605–649. doi: 10.1007/BF00992905

Tomelleri, S. (2010) A Sociology of Emotion: Resentment. *Sociology Studies* [Studi di sociologia], 48, 3–15.

Turnaturi, G. (Ed.). (1995) *The Sociology of Emotions*. Milan: Anabasis Editore.

Van Laer, J. and Van Aelst, P. (2009) Cyber-Protest and Civil Society: The Internet and Action Repertoires of Social Movements. In Y. Jewkes and M. Yar (Eds.), *Handbook on Internet Crime*, (pp. 230- 254), Universia Press, PortlandVincent, J. and Fortunati, L. (Eds.). (2009) *Electronic Emotion: The Mediation of Emotion via Information and Communication Technologies*. Oxford: Peter Lang.

Yang, G. (2000) Achieving emotions in collective action: Emotional processes and movement mobilization in the 1989 Chinese sutfent movement. The Sociological Quarterly, 41 (4), 593–614

11 Cosmopolitan Empathy and User-Generated Disaster Appeal Videos on YouTube

Mervi Pantti and Minttu Tikka

INTRODUCTION

Today's humanitarianism and media are inextricably linked in a co-dependent, mutually beneficial relationship. The media is critical to the humanitarian response of global disasters as we respond to them through media images and discourses that can invest them with emotional and political charge, needed for mobilization of solidarities (Benthall 1993: 8; Pantti et al. 2012; Tester 2001). New computer-mediated technologies are now transforming the communicative conditions of humanitarian-media field, and at the same time reconfiguring communications power and emotional politics in times of disaster.

One of the communicative changes in the humanitarian-media field concerns the increased participation of ordinary people in disaster communications—and the accompanying proliferation of public emotions on- and off-line. On the one hand, mainstream news media have increasingly integrated citizen eyewitness visuals and accounts into their reporting that are appreciated both by media professionals and audiences for adding to the sense of closeness and 'emotional realism' (Pantti and Bakker 2009; Williams et al. 2011). On the other hand, we have seen an emergence of self-organizing 'digital volunteers' who participate in disaster response efforts in a number of ways. This chapter discusses one of the so-far overlooked aspects of citizen communications taking place today in the aftermath of disasters, user-generated videos on the Internet that aim to elicit compassion and raise funds for disaster victims.

Taking as a starting point the concept of 'cosmopolitan empathy' coined by Ulrich Beck (2006: 7), understood as an extended capacity to imagine and empathize with the suffering of others beyond one's immediate existence, this chapter engages with the question of: how do user-generated humanitarian appeals call upon to feel compassion toward disaster victims and shape the emotional politics of humanitarian aid? Specifically, it will examine the emotional mobilization and affective regimes of user-generated videos uploaded on popular video-sharing website YouTube in response to the Japan earthquake and tsunami in March 2011 and East Africa's draught and famine.

Popular online spaces such as the video platform YouTube provide a platform for self-expression and, increasingly, for political communication and participation. We set out to study YouTube disaster appeal videos as an example of mediated political participation, an 'actually existing cosmopolitanism' (Robbins 1998), that is located in popular discourses. Cosmopolitanism is understood here as enacted in the mediated emotional expressions and communicative strategies of 'ordinary people' that aim to promote humanitarian action. New media technologies, apparently, have played a crucial role in enforcing a globalization of emotions as they challenge the nation as the ultimate moral community and its monopoly over public emotions. This is important since the cultivation of compassion has traditionally been restricted by the nationally inflected narratives of disasters (Chouliaraki 2006). The grassroots humanitarian appeals on YouTube, then, are conceptualized as forming a global political space of emotional engagements, a space for expressing and evoking emotions and potentially widening the horizon of cosmopolitan imagination.

ORDINARY PEOPLE AND DISASTER COMMUNICATIONS

In a media-saturated global era, as Paul Frosh and Amit Pinchevski (2009) stated, crises and disasters are no longer interruptions to our lives, but have become a routine, experiential ground of them. According to them, media witnessing incorporates audiences into a system of permanent "crisis-readiness," which helps to shape a global sentiment of shared human vulnerability. Consequently, we need to understand crises and disasters today not only as something that media audiences 'see' in the mass media, but as something "they are increasingly socialized to create" both as the ultimate witnesses of events and as *the central producers of mediated testimonies* (Frosh and Pinchevski 2009: 300, emphasis ours). Social networking sites and online news sites offer new opportunities for emotional self-expression and engagement for people who have in some way been touched by a disaster and who need to publicly talk about their feelings or acknowledge the suffering of others. As Shayne Bowman and Chris Willis wrote in their citizen journalism manifesto *We Media* (2003): "The response on the Internet gave rise to a new proliferation of 'do-it-yourself journalism' [. . .] everything from eyewitness accounts and photo galleries to commentary and personal storytelling emerged to help people collectively grasp the confusion, anger and loss felt in the wake of the tragedy" (pp. 7–8).

This proliferation of ordinary experiences and emotions in times of disaster can be seen in terms of the larger 'demotic turn' (Turner 2010), which describes the ordinary individuals' unprecedented possibilities for media visibility and participation—whether or not serving democratic purposes. Also, we can approach it in terms of the emotional public sphere in which emotional expressions are central to political discussion (Lunt and Stenner 2005). At this cultural moment, we are increasingly displaying and reflecting on our emotions in, through, or by the media, individual emotional experience having become a new regime of truth and a basis of self-representation (e.g. Dovey 2000; van Zoonen 2012). Certainly, questions have been raised about the motives and moral implications of public expressions of emotion. For Chouliaraki (2006: 212), emotional self-expression encourages self-absorption that in turn distracts us from social and political engagement rather than extending the boundaries of collective care and moral community.

The argument we want to pursue here is that ordinary people (as non-professionals) are now increasingly appearing as independent agents of humanitarian communication and function as intermediary moral educators for the emotional response (cf. Pantti et al. 2012). Through new media forms they take on humanitarian tasks such as fundraising, which previously belonged to official respondents. While spontaneous volunteerism is not a novel feature in the aftermath of disasters, new information technologies have supported new forms of volunteerism that are possible to all who have a virtual presence and computing know-how (Hughes et al. 2008; Liu et al. 2008; Starbird and Palen 2011). The trend toward new forms of citizen participation in disasters—such as translating and mapping requests for assistance—was commented on by the Chairman of the United Nations Foundation, Ted Turner, in the context of the Haiti earthquake: "Powered by cloud-, crowd-, and SMS-based technologies, individuals can now engage in disaster response at an unprecedented level. Traditional relief organizations, volunteers, and affected communities alike can, when working together, provide, aggregate and analyze information that speeds, targets and improves humanitarian relief" (United Nations Foundation 2011: 7).

The forms of citizen-driven disaster communications described by Turner and examined in the literature are fundamentally different from the user-generated appeals for donations on YouTube in that they facilitate the information flow between affected communities, humanitarian non-governmental organization (NGOs), and digital volunteers (and mainstream media organization reporting on crisis), whereas the YouTube fundraising appeals are targeted at the uninvolved public, at *anyone* subscribing to these YouTube channels. Thus, the communicative situation between the viewers and user-generated disaster appeal videos is comparable to that of the mainstream media reporting on disasters since it involves those who represent the suffering and those who watch from a distance. Despite the new communicative environment, the YouTube appeals, therefore, face the same challenge as traditional media reports or humanitarian agencies' campaigns: how to incite emotion that mobilizes moral action on behalf of those suffering. To help to understand this widely-discussed problem involved in humanitarian appeals we next shortly consider emotional and ethical changes in humanitarian narratives of the fundraising appeals that construct human suffering.

HUMANITARIAN APPEALS AND EMOTION

Emotional representations, which are increasingly visual-based, are the principal means that humanitarian agents of all kinds have to shape our engagement with the distant sufferer and educate us about how we should respond. However, the paradox of humanitarian representation is that while they call upon us to respond to the suffering in terms of moral action, the witness has, as Luc Boltanski argued (1999: 17–19, 149–169) only two options for action: paying and speaking. Paying as an action in response to a distant suffering means simply giving money to relief aid. Speaking refers to the role of the spectator as a witness of suffering. However, speech according to Boltanski has to meet certain criteria in order to qualify as moral response: it needs to be intentional, committed, and public. It is partly because of the lack of possibility for direct action that the Western public is prone to reject the moral demand (about audiences' denial, see Seu 2010). Audiences' unresponsiveness has also been explained to be due to patterns of mass media coverage. For example, representations of suffering have been repeatedly critiqued for failing to bridge the gap between Western audience and the distant sufferer: "The representation of the suffering that goes together with a portrayal of humanitarian action, especially on television, is intrinsically bad because it transforms the spectator into a voyeur, stimulating his perverse desire to take pleasure in the suffering of others or, at best, provoking feelings of shame for not being able to assuage the suffering that is being shown" (Boltanski 2000: 5).

Humanitarian representations are culturally bound and studies on fundraising campaigns have detected noticeable changes in humanitarian organization's emotional styles of appealing. Lilie Chouliaraki (2010) claims for a trajectory from 'shock effect' campaigns from the 1960s to mid-1980s that focused on the suffering of victims and aimed at inciting guilt, shame, and indignation, via 'positive imagery' campaigns from the mid-1980s to the early-2000s, which changed the focus from the sufferer's victimhood to his personalized agency and operated in the emotional regime of tender feelings such as empathy and gratitude, to the contemporary 'post-humanitarian' styles of appealing. Chouliaraki (2010) sees this emerging 'post-humanitarian' style of appealing as an attempt to renew the legitimacy of humanitarian campaigners, to break through audiences' lack of engagement, and adjust to the media environment in which humanitarian organizations operate today. The shift toward 'post-humanitarian' representational strategies is strongly connected to humanitarian organizations' contemporary increased engagement with the logic of marketing and consumerism (Chouliaraki 2010; Musarò 2011; Vestergaard 2008).

The 'post-humanitarian' style of appealing departs from the earlier campaigns both in terms of representational practices and its moral foundation. While traditional aid appeals relied on the moral universalism of the audience, on the assumption that the audience will feel for the sufferer, the key feature of 'post-humanitarian' appeals lies in abandoning 'grand emotions' and moral universalism as a justification for public action and, instead, relying on the individual judgement and low-intensity emotional regimes (Chouliaraki 2010). 'Post-humanitarian' appeals are characterized by 'technologization of action' and 'de-emotionalization of the cause.' The first means that the use of the Internet has simplified the donating and engagement with the humanitarian cause. The second imperative of the style of post-humanitarian appeal is the avoidance of traditional emotion elicitations (guilt, indignation, empathy) considered not to lead to action (Chouliaraki 2010). This corresponds to a shift in humanitarian narratives from depicting suffering to inviting reflexivity over ourselves and our moral agency (Vestergaard 2010).

These debates about the moral response toward suffering and transformations of humanitarian communication have relevance for our discussion, because they provide a context in which to evaluate the non-institutional humanitarian appeals made possible by the Internet. In the light of them it will be interesting to consider how the videos that appear on YouTube engage emotions and make claims to act on suffering. We have selected 12 user-generated videos which appeal for help to the Japan disaster and the East Africa famine victims. We have divided these videos into two categories according to their form: vlogs (video blogging consisting of individual speaking to camera about his or her reaction to one of these two disasters), video montages (video consisting of self-made, or existing footage, pictures, images, words, and sound, combined into a new text). The methodology involves analyzing the audio-visual elements (images, speech, music, sound effects) and textual elements of the videos (title, subtitle, description of the video with links, written comments upon video). We have made observations about emotional expressions, affective regimes acted out in the narratives that connect viewer and sufferer and justifications of action in these videos. We will first look at the user-generated appeals for the Japan disaster and then compare and contrast the findings with the appeals for the African Horn famine. Finally, the discussion will return to questions of cosmopolitan empathy in a participatory culture of YouTube.

VLOGGING JAPAN: HELP *ME* TO HELP

The format of a vlog is a typical form of participation on YouTube, where the maker of the video films is her/himself talking and commenting on something. A distinctive feature of vlogging is its conversational, intimate character that is not unlike interpersonal face-to-face communication (Burgess and Green 2009: 54; Tolson 2010). All the appeal videos we examined come from individuals who are regular vloggers: some are witnessing the disaster at a distance, some from within Japan.

The disaster appeal videos can be conceptualized as mediated testimonies: the YouTube users bear witness before the camera by documenting what they have experienced mainly through media. At the center of vlogs is the expression of emotions of the vloggers themselves—often sadness, empathy for victims, and most of all, desire to help. The suffering of victims is present only as the object of the vloggers' feelings. The connection between the viewers and the victims of the disaster, together with the responsibility to participate in the relief response, is produced solely by the experience of the vlogger, and it becomes a justification for asking the implicit viewer to share the moral responsibility. For example, in a vlog entitled "Help me HELP JAPAN,"[1] lasting about eight and half minutes and recorded as a single close-up shot, Jason who is living in Japan but is visiting the United States starts his story by talking about how he has been searching for information from television, newspapers, and social media about the disaster area. He repeatedly tells how bad and helpless he feels for not being able to help and comfort the people there. "I want to do something positive," he says, and promises to pay 50 cents for each positive comment on Japan posted and donate the money to the Japanese Red Cross.

Besides from their sincere expressions of emotions and willingness to help, the emotional identification with the vloggers arises from the aesthetic style of vlogs, which is characterized by the direct address, the everyday language, the private mise en scène, and the amateurishness of the production, all of which add to the sense of authenticity of these appeals. The sense of authenticity is reinforced by the ordinariness of the campaigners, which is seen, for example, in simple and straightforward, almost childlike appeals for donations such as "please help if you can."

This ordinariness as a basis for emotional appeal and trust is most pronounced in Whiteboy7thst's video entitled "HELP JAPAN!!!"[2] in which he characterizes himself as someone who "doesn't go to school" and staggers on words when reading aloud text from the American Red Cross website. Like most of the vloggers, he offers as a justification for his appeal having seen images of destruction in the media that "really hit home." His intensively felt personal experience led to the need to raise awareness about the disaster among his subscribers (gamers) whose demographic is young people between 12 and 21 and who are not informed "as they do not follow news."

That the vlogs rely on the authenticity of emotions and extreme intimacy created by the testimony of the vlogger to construct the humanitarian cause is shown also in the vlog appeal titled "Please help Japan!!! Japan Earthquake"[3] posted by a young Japanese woman who usually vlogs about fashion and beauty. The video is created as a single, unedited shot of her face, directly addressing the viewer as she appeals for donations and asks people to share information about donation sites. She starts her serious monologue conversationally by referring to disaster as "news I didn't want to hear about" and thanks her followers for being concerned about her well-being. Next, her voice breaks when she tells about not having heard from a friend living in the disaster area. The vlog enacts a sense of global solidarity through presenting her personal emotions as feelings all viewers share: "Japan we all love." Here is a transcript of the video's English translation:

> March 12th. Since yesterday, there has been news I didn't want to hear about. First of all, thank you so much for worrying about me and my family. All of my family members are safe. However, so far I haven't been able to contact one of my friends who lives near the earthquake zone. I thought I would say more, but the most I want to say is for the people who watch my videos. Please help the people who live in Japan, foreigners in Japan, and all the people. That is all I wanted to tell you.
>
> I want all people to watch and understand my videos, but I don't know English that much. Someone please translate if you can. The damage is unimaginable serious, and there are many people out there who are still suffering. It has been over 24 hours since the earthquake and collection of donations for the areas that suffered terrible damage is underway at a rapid pace within Japan. I'd like to use this video to gather together all information on donation activities, both within Japan and across the worlds. If you have any information about donations for Japan, or any URLs etc, please leave a message in the comment section below. Let's share this information.
>
> However, it's a fact that donation fraud sometimes occurs. I ask everyone please check carefully, and then lend support to an authentic fund-raising body. With the support of the people of the world, I hope we are able to aid the recovery of the Japan we all love. Please help if you can.

An interesting question is how the ordinariness of the makers of these appeals affects the reception, given that the presence of "ordinary people" in the media is claimed to lay audiences' expectations on the expression of the 'truth' beyond the processes of mediation (e.g., Livingstone and Lunt 1994). In her study on audiences' denial of humanitarian appeals, Irene Seu (2011: 453) argues that members of the audience position themselves as victims of NGOs' sleek and manipulative appeals. Consequently, she argues, the ways in which the sufferer is represented (e.g., whether the sufferer is attributed agency or characteristics audiences can identify with) matter less than audiences' relationship with those who appeal to them (p. 454). What she finds, then, is a deep distrust in both NGOs' commercialized, brand-oriented practices and in the media in general. While the peer-to-peer nature of amateur humanitarian appeals together with the visual and textual elements that communicate authenticity may help to circumvent the 'shoot the messenger' effect, the wider discourse of distrust is very much present in this amateur video sphere, operating within a broader media context. It is seen, for example, in how the makers of appeals emphasize the authenticity of NGOs they are raising funds for (all YouTube videos promote different official humanitarian agencies), as seen in the appeal of the Japanese woman earlier, or in how they try to safeguard their trustworthiness by showing evidence of their own transaction for an aid agency.

The last point we want to make concerns the moral agency of the vloggers and their viewers. The Japanese woman requests her viewers translate her report into English and post information about the websites of humanitarian relief providers. Besides English, her video "Please help Japan!!!" has been translated into French, Italian, and Chinese, and in her video description there are over 30 options available for donations to Japan disaster relief. Jason appeals to his subscribers to participate by posting positive comments on Japan, such as "I have never been to Japan but always wanted to see a live sumo match." These examples show that the vloggers, and YouTube appeals in general, request help not only through donations but attempt to mobilize response also in terms of 'speaking' on behalf of the victims, which according to Boltanski (1999) offers a political opportunity to step out of the passive role of the spectator (if it is done with the intention to alleviate suffering), unlike the mere paying, which makes a limited difference in the reality of the donor.

CELEBRATING *OUR* SOLIDARITY

While vlogs call for identifying with the anxiety and empathy of the makers of appeals, the video montages aiming to raise funds for the Japan disaster appeal to 'common humanity' and rely on eliciting feelings of solidarity among their audiences. The intimate 'you' addressed in vlogs, then, becomes an empowering 'we' in video montages: it is the need to realize our own humanity, rather than others' suffering, that is offered as a reason for action.

Video montages that cut and mix various forms of existing media content and/or record one's own video material rely on different textual strategies from animation to a pastiche of photographs in making their case. The video montages are profoundly inspired by popular culture and celebrity, and the emotional tone of them, in contrast to vlogs that are sombre and un-ironic, is predominantly playful and joyful. A graphic animation called "Support for Japan: Godzilla can not do it alone"[4] aims to provoke emotional reactions and action with the help of an iconic Japanese movie monster. The scenes in which Godzilla destroys Tokyo are played in reverse so that it looks like the monster is rebuilding the devastated city. The only text of the appeal "Godzilla can not do it alone" appears in the end of the clip accompanied by the emblem of the Japanese Red Cross and indisputably calls for cosmopolitan solidarity.

Scholars have pointed out that new humanitarian discourse is focused on the self-identity and values of the donor: it "argues for humanitarian action on the basis of an ethical capital with which it provides the donor" (Vestergaard 2010: 177). Consequently, the suffering of others mainly remains absent and the focus is on the similarity of humanity. In appeal videos, 'we' as the potential donors are represented primarily in terms of gratifying emotions: the message is that solidarity makes us feel better about ourselves. Consider a video called "JAPAN DISASTER Pray for Japan . . . from NewYork"[5] This video appeal is a pastiche of pictures of ordinary people from different ethnic backgrounds photographed on the streets of New York. The soundtrack of the video consists of sentimental songs of Michael Jackson and Aerosmith. People are holding a big notepad where they have written their greetings to Japanese people such as "Recover soon!," "I think of you," and "God bless Japan."

Some of the video appeals are constructed as entertainment in which the celebrity of the appeal maker or a person unrelated to the disaster is used to attract attention. An example of an appeal video that relies on the celebrity of the YouTube maker is called "Honk for Japan!"[6] The producer of the video is an Internet star, Ryan Higa, known as 'Nigahiga' on YouTube, who makes comedy videos. On March 14, 2011, he uploaded the video "Honk for Japan!" that has been viewed at the time of writing about 8.5 million times. 'Nigahiga' starts his video by addressing the audience directly in edited close-up shots: he wants to help Japan and to encourage others to help as well. He promises to donate $10 for every honk he gets to Red Cross's disaster relief fund for Japan (so far he has donated $6,475) and writes at the end of his description, "You can help support as well by making a donation at: http://www.REDCROSS.org." Next, we see him beside an intersection in Los Angeles holding a sign saying "Honk if you love Japan" and having fun with his friends as people drive by and honk. This YouTube celebrity is actively employing his popularity to reach the viewers as he tells in the video description[7]:

> My biggest strength right now is that I'm fortunate enough to have a following. The purpose of this video was not only to support Japan myself, but to encourage others as well. I could have easily donated the money and not made a video about it, but I think it's a lot more important to get support from all over the world.

A very different celebrity-driven appeal calling for self-reflection about responsibility is titled "Pray by Justin Bieber: Pray for Japan."[8] The appeal video is made by a fan of the American teen idol Justin Bieber, and unlike the other montage appeals, it operates at the register of the tragic. This video mixes written text and high-quality news stills of the Japan disaster showing destruction, suffering, and people praying. The sad and solemn tone of the appeal is reinforced by Bieber's song "Pray (I Close My Eyes)," that is used as a soundtrack. The clip opens with three still images picturing sushi, Japanese artwork, and a painting of Japanese women wearing kimonos with overlaying texts: "Some of us like their food. . . . Some of us like their art. . . . Some of us like their culture." Next, we see a black screen with a text: "Where is everybody when they need our help?" The clip ends with a red screen with the text "Click the link in description to donate and help rebuild Japan." Here, the audience is asked to identify as moral actors with the suffering of Japanese people as the video poignantly delivers a message of their humanity as a shared humanity. While it emphasizes the similarity of humanity, it differs from other user-generated appeals for Japan by actually depicting suffering—even if in a sublime form—and eliciting not only empathy toward the victims but also guilt and shame to motivate moral behavior ("Where is everybody when they need our help?").

We will next discuss Horn of Africa famine appeals on YouTube, which unlike the Japan appeals, aim to raise funds by focusing on the suffering of the famine victims and operate in the traditional affective regimes of humanitarian communication.

WITNESSING SUFFERING IN AFRICA

The appeal videos aiming to raise funds for the East African famine noticeably differ from the appeals for the Japan disaster in several ways. First, the amount of user-generated videos on the East African famine is very low compared to the amount of videos on Japan's disaster. Most videos that appeal for help to East Africa uploaded on YouTube are made by official humanitarian organizations, not by ordinary users. Second, the user-generated videos on East Africa have low view counts and only a few comments. To compare, in the case of Japan the amount of views is between hundreds of thousands and millions, while in the case of East Africa the amount of views ranges from a thousand to ten thousand. Third, while the Japan appeals were focused on the emotions of the maker of the video or those of the wider (Western) audience, the video montages regarding East Africa bring the sufferers to the center of the imagery. As humanitarian narratives they encourage the audience to act both out of an empathetic connection with the sufferer, and out of responsibility toward the suffering. Thus, they predominantly draw on the 'old' humanitarian discourses by employing the affective register of compassion, guilt, and shame. Fourth and finally, the narrative strategies through which the videos seek to mobilize emotions do not represent the same range of variety as video appeals for Japan but rely on photorealism.

A video entitled "Horn of Africa drought 2011 HELP TODAY"[9] illustrates the mobilization of emotions through conventional tragic and horrendous stories. The video begins with dramatic music by an African singer Salif Keita and a map that illustrates the situation in the Horn of Africa, namely, the fact that the area is experiencing its worst draught in 60 years, and that 12 million people are estimated to be affected. The video continues with a montage of 22 documentary pictures of the emerging disaster: dried-up soil, a desperate mother carrying her emaciated child, children standing in a line with rusty bowls, a mother giving water to a baby, a man kneeling down beside a dead camel, and a child whose ribs are exposed lying on a bench. These pictures represent the horrendous reality of suffering, and the eye contact in many of these pictures establishes a relationship between the spectator and the sufferer. Most of the pictures consist of a mother–child composition, or children with their water bowls. The pictures of the suffering women and emaciated children represent the traditional iconography of distant suffering. The video ends with four short consecutive statements: "Think & Ask yourself ... How can I help ... Donate ... And pray immensely." The description offers further justification for participation by explaining the complexity of the situation from the viewpoint of the UN and NGOs.

All of the user-generated videos on East Africa utilize the imagery and rhetoric of the early, 'shock effect' humanitarian campaigns, which have been criticized for objectifying the sufferer and increasing the distance between 'us' and 'them.' The only exception is the video named "East Africa Appeal,"[10] which combines documenting others' suffering with a 'frame story' that is focused on Western solidarity and awareness. The video was partly filmed at the Summer in the City Gathering 2011, the YouTube community's annual meeting in London. The video begins with edited clips of the participants sitting in a park and talking about East Africa's situation using everyday language. The video continues in East Africa and presents "Umi's story," a clip about a 3-month-old baby suffering from dehydration and starvation. After this, the video returns to the park where the participants ask the viewers to help "children like Umi to survive." It ends with a group picture of the participants smiling and shouting "Thank you!" The appeal, then, combines traditional 'negative imagery' that aims at inciting emotional responses of compassion in viewers with contemporary, donor-centered strategies of humanitarian communication which offer humanitarian action as a joyful and rewarding experience.

CONCLUSION

It has been said that YouTube offers a new site of "cosmopolitan cultural citizenship" (Burgess and Green 2009: 79) because of its rootedness in ordinary experiences and enhanced opportunities for participation and public dialogue, *partly* autonomous from commercial media (Strangelove 2010). As a result of the mediatization and technologization of witnessing suffering, there are new opportunities for the expression and the enactment of emotion which, in turn, make possible the formation of 'feeling communities' and acting on a humanitarian cause outside of the framework of and roles given for ordinary people by traditional media and humanitarian organizations—and beyond national boundaries. As we stated, at this juncture to witness disasters means more than to just 'see' them in the media—it is increasingly about producing personal testimonies on suffering, as a response to media representations of suffering.

YouTube video appeals, entering in the field of fundraising that has belonged to the institutional players, represent a new form of humanitarian communication—an emotional public sphere in which collective and individual emotions are expressed, knowledge is shared, and political issues are discussed. The makers of the YouTube video appeals often combine the two options the media witness has for action, paying and speaking, and at the same aim to mobilize collective action by encouraging viewers to do the same. Thus, the appeal videos aim to move their viewers beyond being merely 'feelers' or donors to be feeling participants.

YouTube video appeals, then, exemplify the potential of popular as a resource for political participation, adding different forms of expression and emotional registers to the institutional forms of humanitarian communication. The emotional logic of YouTube appeals is not entirely corresponding with the low-intensity emotional regimes and reflexivity of 'post-humanitarian' appeals even if "the 'necessary' link between seeing suffering and feeling for the sufferer" is absent in many YouTube appeals (Chouliaraki 2010: 119). Drawing on a variety of textual strategies and emotional registers, all video appeals straightforwardly ask us to feel responsible for and act on suffering outside of our own life world.

While ordinary people's appeals for donations clearly illustrate the expansion of the range of voices involved in disaster communication in the present media environment, these voices, however, do not challenge the legitimacy of relief agencies but rather act as a go-between for them and the public. This grassroots emotional education may help, to some extent, to get around the widely-documented distrust in humanitarian organizations (and in mainstream media), or at least reach unlikely donors, young people in particular. The impact of these user-generated videos is difficult to verify in terms of donations or further acts toward the humanitarian cause. The official NGOs are today actively using new information technologies and social networking platforms as new avenues for disseminating information and raising awareness and funds (e.g., Cottle 2009: 160–163). The International Red Cross and its many national associations have uploaded over 30 videos on YouTube relating to Japan and East Africa disasters. However, their view counts and the number of comments are significantly lower than those of user-generated videos.

What these appeal videos from ordinary people also show is that the emotion expressed and evoked in YouTube are no less connected to global relations of structural inequality than mainstream media representations. The fact that the cosmopolitan empathy enacted in user-generated videos is rooted in the emotional experiences and life worlds of YouTubers means that culturally distant sufferers are accorded much less attention than more proximate ones. It is also seen in that appeal videos for East African famine victims use more traditional representational practices to prompt emotional responses. This shows that studying YouTube videos as emotional texts that create objects and subjects of feelings as well as communities of feelings must necessarily be done in relation to other humanitarian discourses and media technologies.

NOTES

1. http://www.youtube.com/watch?v=Zghh9kNnIRc. Retrieved on November 9, 2011.
2. http://www.youtube.com/watch?v=KSWTG2Z0Sj8. Retrieved on July 18, 2013.
3. http://www.youtube.com/watch?v=TIe2sdJTy-Y. Retrieved on July 18, 2013.
4. http://www.youtube.com/watch?v=jrIFX0BaHcU. Retrieved on July 18, 2013.
5. http://www.youtube.com/watch?v=JPBMexnWXgI. Retrieved on July 18, 2013.
6. http://www.youtube.com/watch?v=cciUXpITsu0. Retrieved on July 18, 2013. The video has currently 9 628,777comments, and about 240,000 'likes' and 2,800 'dislikes.'
7. NigaHiga's subscribers have taken the idea and produced their own 'Honk If You Love Japan' video campaigns. See e.g., http://www.youtube.com/watch?v=eBATcfB-rtY. Retrieved on July 18, 2013.
8. http://www.youtube.com/watch?v=sGl67zM45U4. Retrieved on November 11, 2011.
9. http://www.youtube.com/watch?v=GpgQq_hTjE4. Retrieved on July 18, 2013.
10. http://www.youtube.com/watch?v=WB7vDOU8i8M. Retrieved on July 18, 2013.

REFERENCES

Beck, U. (2006) *The Cosmopolitan Vision*. Cambridge: Polity Press.
Benthall, J. (1993) *Disasters, Relief and the Media*. London: I.B. Tauris.
Boltanski, L. (1999) *Distant Suffering. Politics, Morality and the Media*. Cambridge: Cambridge University Press.
Boltanski, L. (2000) The Legitimacy of Humanitarian Actions and Their Media Representations: The Case of France. *Ethical Perspectives*, 7(1), 3–16.
Bowman, S. and Willis, C. (2003) *We Media: How Audiences are Shaping the Future of News and Information*. Reston, VA: The Media Center at the American Press Institute. Retrieved on 18 July, 2013, from http://www.hypergene.net/wemedia/download/we_media.pdf
Burgess, J. and Green, J. (2009) *YouTube. Online Video and Participatory Culture*. Cambridge: Polity Press.
Chouliaraki, L. (2006) *The Spectatorship of Suffering*. London: Sage Publications.

Chouliaraki, L. (2010) Post-Humanitarianism: Humanitarian Communication: Beyond a Politics of Pity. *International Journal of Cultural Studies*, 13(2), 107–126.
Cottle, S. (2009) *Global Crisis Reporting: Journalism in the Global Age*. Maidenhead: Open University Press.
Dovey, J. (2000) *Freakshow: First Person Media and Factual Television*. London and Sterling: Pluto Press.
Frosh, P. and Pinchevski, A. (2009) Crisis-Readiness and Media Witnessing. *The Communication Review*, 12(3), 295–304.
Hughes, A., Palen, L., Sutton, J., Liu, S. and Vieweg, S. (2008) Site-Seeing in Disasters: An Examination of On-Line Social Convergence. *Proceedings of Information Systems for Crisis Response and Management Conference (ISCRAM)*. Retrieved on July 18, 2013, from http://www.cs.colorado.edu/~palen/Papers/iscram08/OnlineConvergenceISCRAM08.pdf
Livingstone, S. and Lunt, P. (1994) Talk on Television: Audience participation and public debate. London: Routledge.
Liu, S., Palen, L., Sutton, J., Hughes, A. and Vieweg, S. (2008) In Search of the Bigger Picture: The Emergent Role of On-Line Photo-Sharing in Times of Disaster. *Proceedings of Information Systems for Crisis Response and Management Conference ISCRAM 2008*. Retrieved on July 18, 2013, from http://www.cs.colorado.edu/~palen/Papers/iscram08/OnlinePhotoSharingISCRAM08.pdf
Lunt, P. and Stenner P. (2005) The Jerry Springer Show as an Emotional Public Sphere. *Media, Culture and Society*, 27(1), 59–82.
Musarò, P. (2011) Living in Emergency: Humanitarian Images and the Inequality of Lives, *New Cultural Frontiers*, 2, 13–43. Retrieved on July 18, 2013 from http://www.newculturalfrontiers.org/wp-content/uploads/New_Cultural_Frontiers_2_2_Musaro%CC%80.pdf
Pantti, M. and Bakker, P. (2009) Misfortunes, Sunsets and Memories: Non-Professional Images in Dutch News Media. *International Journal of Cultural Studies*, 12(5), 471–489.
Pantti, M., Wahl-Jorgensen, K. and Cottle, S. (2012) *Disasters and the Media*. New York and London: Peter Lang.
Robbins, B. (1998) Introduction Part I: Actually Existing Cosmopolitanism. In P. Cheah and B. Robbins (Eds.), *Cosmopolitics: Thinking and feeling beyond the nation* (pp. 1–19). Minneapolis: University of Minnesota Press.
Seu, I. (2010) "Doing Denial": Audience Reaction to Human Rights Appeals. *Discourse & Society*, 21(4), 438–457.
Starbird, K. and Palen, L. (2011) "Voluntweeters": Self-Organizing by Digital Volunteers in Times of Crisis. In *Proceedings of CHI' 2011*. Retrieved on July 18, 2013, from http://www.humanityroad.org/voluntweetersStarbirdPalen.pdf
Strangelove, M. (2010) *Watching YouTube: Extraordinary Videos by Ordinary People*. Toronto: University of Toronto Press.
Tester, K. (2001) *Compassion, Morality and the Media*. Buckingham: Open University Press.
Tolson, A. (2010) A New Authenticity? Communicative Practices on YouTube. *Critical Discourse Studies*, 7(4), 277–289.
Turner, G. (2010) *Ordinary People and the Media: The Demotic Turn*. London: Sage.
UN Foundation. (2011) *Disaster Relief 2.0: Information Sharing in Humanitarian Emergencies*. Retrieved on July 17, 2013, from http://www.unfoundation.org/disaster-report.
Vestergaard, A. (2008) Humanitarian Branding and the Media: The Case of Amnesty International. *Journal of Language and Politics*, 7(3), 471–493.

Vestergaard, A. (2010) *Distance and Suffering. Humanitarian Discourse in the Age of Mediatization*. Unpublished PhD thesis. Copenhagen Business School. Retrieved on July 18, 2013, from http://openarchive.cbs.dk//bitstream/handle/10398/8318/Anne_Vestergaard.pdf?sequence=1

Williams, A., Wahl-Jorgensen, K. and Wardle, C. (2011) "More Real and Less Packaged": Audience Discourse on Amateur News Content and Its Effects on Journalism Practice. In K. Andén-Papadopoulos and M. Pantti (Eds.), *Amateur Images and Global News*. Bristol: Intellect (pp. 193–210).

van Zoonen, L. (2012) *I*-Pistemology: Changing Truth Claims in Popular and Political Culture. *European Journal of Communication*, 27(1), 56–67.

12 Anger, Pain, Shame, and Cyber-Voyeurism
Emotions Around E-Tragic Events

Alessandra Micallizzi

The topic of emotions has recently received renewed attention from social scientists. Though generally considered more pertinent to psychology, emotions have been the object of study by sociologists. Recent international events confirm the importance of examining the social aspects of the practice of sharing emotions.[1]

This chapter presents both theoretical and empirical evidence about the characteristics and the dynamics of sharing emotions on the net, which is considered a special emotional social context. The term "cyber-voyeurism" refers to the phenomenon of observing and following private tragic events that have been made public by virtue of re-publication of news content on the Internet. This phenomenon is interesting for two reasons: it is rich in emotional components, and even if it starts in the traditional media, it generally takes place in participative spaces on the net, typically on social network sites.

The first section of this chapter offers a description of the main theories on the practice of sharing emotions and its consequences, and on the net as social and narrative context for discussing, coping with, and socializing emotions.

The second section deals with the empirical study of the conversations posted on YouTube as comments to an amateur video dedicated to Sarah Scazzi, a teenager who disappeared in August 2010 and was found dead a few months later. Almost all the people who posted a message on the YouTube video had never met Sarah, but their words evidence a strong emotional involvement with her story. Based on findings arising from the content analysis and conversational analysis, I shall discuss the dynamics of social sharing of emotions and the characteristics of the sentiments expressed.

In the conclusion, an interpretative model of cyber-voyeurism is proposed, along with the possible reasons for its dissemination as social behavior.

SHARING EMOTIONS AS CYBER-PRACTICE

The theoretical framework supporting this empirical study comprises various studies on the social sharing of emotion, as proposed by Rimé (1995,

2009). Taking this as my point of departure, I shall reflect on the role of media events in producing cognitive ruminations and stimulating the practice of sharing emotions. I shall furthermore discuss the net as a narrative and emotional social context, in order to introduce the hypothesis of a cyber-social sharing of emotions.

Emotive Implications of Traumatic Experiences

The affective life of individuals has important psychological implications, but is also closely connected with the social context. The expression of emotions is governed by strict social rules and cultural norms. Sociological studies have examined, among others, the role of emotions in interpersonal and group relations, the linkage between feelings and the social conception of the self, and the power of emotions in societal processes of stability and change (Kemper 2000).

In this chapter, attention is focused on the social sharing of emotion, and in particular on the so-called "empathic role-taking" sentiments (Shott 1978). Shott coined this term to define emotions that provide a vicarious emotional experience, i.e., putting oneself in someone else's shoes. Beside this class of emotions, there are the "reflexive role-taking" emotions, which include guilt, shame, and embarrassment. Shott's theory is interesting in its description of the role played by emotions in relation to others. In fact, we can say that studying the social sharing of emotions means considering the roles of others in the processing of personal emotive experiences, especially negative ones.

Empirical studies have confirmed that exposure to an emotional situation entails social consequences (Rimé 1995). One of the more widely studied aspects is the tendency to seek contact with others in situations of distress (Gump and Kulik 1997). Furthermore, experimental evidence shows that direct or indirect experience of traumatic events generates a cognitive activity consisting in "a form of intrusive and repetitive thoughts and images related to the emotion-eliciting situation" (Luminet et al. 2000: 663). Starting from these two established points and supporting them with other studies, Rimé (1995) points out that exposure to traumatic situations leads to "social sharing of emotions." This social sharing involves: "(a) the evocation of the emotion in a socially shared language; (b) at least at the symbolic level, an addressee, which occurs when individuals communicate openly with one or more others about the circumstances of emotion-eliciting events and about their own feelings and emotional reactions" (Luminet et al. 2000: 663).

Thus, the social sharing of emotions is based on linguistic sharing of feelings and personal thoughts about the event, and is characterized by the presence of a referent, albeit symbolically. This practice of "translating" emotions into a shared discourse has an *adaptive function*, as its purpose is to cope with the shock caused by the traumatic event (Pennebacker 1997).

According to Pennebacker, the most successful strategy for sharing emotions is expressive narration, i.e., the practice of self-disclosure, especially in writing. As mentioned earlier, *cognitive rumination* (Rimé 2009) can be generated even when the episode does not directly involve the subject, by empathic projection of the self into an external situation.

For the purpose of the present study, it is worth reflecting on the role of media products in capturing the attention of audiences and involving them in mediated emotive experience. As suggested by Peters (2011), news about private tragedies or crimes reminiscent of the spectator's life experiences creates a sense of emotive participation, because "we find ourselves in the position of spectators at a drama without the relief of knowing that the suffering is unreal" (Peters 2011: 39). Moreover, Peters points out that the audience is generally captured by an inexplicable "fascination for trauma without the exoneration of knowing it is all an experiment of mimesis" (2011: 39). This is also one of the conclusions of Rimé's studies (2009) on prudery regarding traumatic stories, exemplified by the typical situation near a car accident: passers-by tend to stay on the scene, fascinated, to capture more details about the event.

If we combine Peters' analysis of media news and Rimé's studies on the sharing of emotions, we can conclude that:

- There is a natural tendency to be fascinated by traumatic and tragic events, whether directly experienced or reported by media news;
- At the same time, traumatic situations entail cognitive rumination consisting in thinking repeatedly about the episode;
- The adaptive response to these intrusive thoughts is the social sharing of emotions.

The Net as a Narrative Cyber-Context

It is generally accepted that the net is not simply a socio-cultural artifact (Flichy 2007), but can be thought of as a narrative social context with its own specific characteristics (Walzer 2001). First of all, while the net favors the trend of using audio-visual languages, the main code used on the net is still textual, and this has some consequences. Whereas the language is light, contracted, and rapid, as in a face-to-face conversation, it is also mediated and reflexive, as in written communication. Ong (1982) famously defined this new form of language as "secondary orality." As boyd (2007) suggested, the net is characterized by scalability, persistence, replicability, and searchability. Thus the mediated communication is not ephemeral, and this has important implications when people share emotional content, as described next.

Mediated communication implies a rethinking of the idea of presence. Although we cannot examine this concept[2] in greater depth here, it is important to underline, on one hand, the sense of togetherness experienced in virtual environment (Bakardjieva 2003), and on the other, the reduced responsibility felt toward others, as suggested by Turkle (2011). Pata (2011) uses the term *hybrid narrative environment* to define the net, in tune with our research, in the sense that the narrative fragments originate from hybrid sources: people, the media, and a process of remediation (Bolter Grusin 2002). At the same time, Pata (2011) recognizes the *narrative nature* of the net as ecological environment. The net can be described as "a processing space of the networked public opinion, where cultural citizenship is experienced with a personal reflexivity, above all through conversational dynamics" (Boccia Artieri 2010: 1). From this point of view, the net becomes a new social context for sharing emotions about traumatic events, combining the power of written self-disclosure (Pennebacker 2004) and the more or less symbolic presence of others.

VOYEURISM AND THE *PRUDERY* OF NET-UPDATING: THE STUDY

In popular culture, the term "voyeurism" refers to a person's tendency, not necessarily of prurient nature, to observe something without the knowledge of others. Voyeurism (from the French *voyeur*, literally "viewer") can take various forms, but typically voyeurs are not directly connected with the person of interest, who is unaware of being observed. The term "voyeur" may also describe a person who enjoys watching the suffering or misfortune of others.[3] The definition thus includes three features: the voyeur watches someone else without being seen; he/she doesn't have a real relation with the observed; the viewer's interest is focused on misfortune or negative events involving others.

If we consider the nature of the net, as described earlier, it is clear that this kind of behavior is perfectly suited to online situations, where nicknames allow people to conceal their real identities, where it is possible to access information about private aspects of unknown others' lives, where news channels use their space to re-mediate specific contents and to produce interactive communication with their networked audience (Kazys 2008). Furthermore, to cite Rimé (2009) once again, negative events generate more interest in viewers, especially when brutal or traumatic, since they cause a cognitive rumination that leads people to empathic projection concerning the event.

Method

An empirical study describing this new socio-emotional practice in greater depth was conducted regarding a crime that occurred in Italy in 2010: the murder of Sarah Scazzi. This personal tragedy, which was originally broadcast on media news, provides an interesting case for examining cyber-voyeurism in view of the following:

- After Sarah's disappearance on August 26, 2010, the most popular hypothesis was that she had left voluntarily with someone she "met" on the net; the Net thus became the focus of a public debate from the very outset of the event;
- Relatives, and in particular Sarah's cousin,[4] used a Facebook page to inform people about Sarah's disappearance and to collect reports and clues on the case;
- More information about Sarah's private life was searched out online, and some clues, such as essays written at school, were circulated on the net even before detectives collected them;
- The boundaries between the public and private spheres were blurred. This became even more evident when Sarah's body was found in real time during a live TV interview with the victim's mother;
- It was a shocking case that provoked a moral and ethical debate, still underway both in the traditional media and on the net.

The present study deals with the dialogue generated by the posting of a video on YouTube by Sarah's best friend four days after Sarah disappeared. Some journalists mentioned the video in their news reports, which resulted in an increased number of visits and comments. The general goal of the study was to describe cyber-voyeurism on the basis of this particular case, and more specifically, to define:

- The narrative characteristics of the posts exchanged on the platform, with particular focus on the emotive expressions and words;
- The main aspects of the dialogues, with special emphasis on the relations between users and their roles;
- The implicit and explicit motives underlying this narrative behavior.

In order to achieve these goals, the Narrative Network Analysis (NNa; Micalizzi 2011) was applied, combining qualitative and quantitative tools. NNa included first of all a qualitative content analysis of the posts, based on the point of view that each comment can be considered a *fragment-de-vie*.[5] Thus, attention was focused on the subject, the object, the referent, and the tone of the post, in a narrative perspective.[6] In addition, a quantitative content analysis of the posts was conducted, using specific software,[7] in order to individuate the emotions described and measure their *narrative weight* in the macro-narration resulting from all the published material.[8] The second step involved the use of conversational analysis in order to define the roles of the main characters in the dialogue. This again was combined with the use of a network analysis software,[9] applied to define the topography of the dialogue. The last step consisted in carrying out in-depth interviews with some people who had posted comments to the video.

The content analysis and the dialogue analysis comprised 3,117 comments, posted from August 30 to November 29. There was much participation during those three months, reflected by the sudden increase of viewings (to 377,714) and the frequent updating of the posts.[10] Identifying the interviewees was complex for two reasons: many of the accounts used to publish comments had been closed, and some of the net users contacted for the interview refused to cooperate and expressed their unwillingness by email. Though this could be seen as a failure to achieve a specific goal of the study, it can also be considered as an important result that confirms some particular characteristics of voyeurism, as will be explained in the last section.

Description of the Results

The first part of this section centers on the qualitative narrative aspects of the posts. Its objectives are:

- To describe the general narrative characteristics of the posts and the dynamics of the dialogue for each phase of the macro-narration. In fact, the qualitative content analysis led to the individuation of four different ways of *living* the Internet as a narrative context, corresponding to four phases of the macro-narration, reflecting four phases in the general story revolving around Sarah's tragedy;
- To identify the main explicit referents of the narration. The narrative analysis and the network analysis allowed me to study the topography of the conversation and identify the referents—where explicit—of the posts analyzed;
- To identify four main profiles of participation, according to the narrative characteristics of the posts published by the protagonists of this space (the YouTube video space).

The second part of this section offers a quantitative content analysis of the narrative material with a specific focus on emotions. Using T-lab and clustering the words around a specific group of emotions, it was possible to calculate the weight of each emotion in the macro-narration derived from all the posts published.

The Qualitative Content Analysis: Narrative Aspects

When examining the macro-narrative level, it is evident that participation was directly connected with some important narrative turning points in Sarah's case. More specifically, it is possible to recognize a change in connotation of the digital space, in line with the evolution of the narration. In fact, the content analysis demonstrates that the micro-hybrid narrative environment (Pata 2011), created on YouTube by the video, was "used" in four different ways and for four different purposes: as a *bulletin board*; as an *investigative office*; as a *courtroom*; and finally, as *a place to share digital memories*.

Anger, Pain, Shame, and Cyber-Voyeurism 199

During the first few days, participation was focused on exchanges of messages, mainly between Sarah's friends, expressing the hope that she would return to Avetrana (her birthplace). Net users described their personal feelings about the situation, expressed emotive support to Brixy94, the video's author, and inquired about clues or signs of Sarah's possible voluntary escape. The tone of the dialogue was very confidential, intimate, similar to that of a dialogue in a familiar public place—such as a coffee shop or a square near home. To illustrate the nature of those exchanges, below are some translated excerpts from posts published during the first days following the video's publication.

> Brixy94 (the video's author): "Sarah will come back! I'm sure . . . please, Sarah . . . if you see this message, call me back. We miss you!"
>
> Giusybella (Sarah's friend): "Please, Sarah, come back, we're really worried about you. We love you."
>
> Mydanilo85 (Sarah's friend): "@Brixy if someone did something to Sarah . . . I don't know what I'd do."
>
> Theje95: "Nice video! @Brixy, nice job! I don't know her, but I hope someone finds her and convinces her to come back home!"
>
> Stefanino10: "Someone saw Sarah at the Palermo train station! Good news guys! Can I have @brixy94 msn contact? I would like to chat with you and give you more details."

The referent is explicit and specific. Some of the messages were written to Sarah, inviting her to return. Most were addressed to Brixy or to other specific net users, tracing a net of the discussion (Figure 12.1).

During this initial phase, some flaming episodes occurred against Celtieg,[11] because he used a provocative tone and made disparaging comments about Sarah (the victim) and Brixy94 (the video's author). The second phase of the dialogue consisted in formulating different hypotheses to explain Sarah's disappearance. These were based on clues proposed by media news, a detailed analysis of the relatives' behavior during public video interviews, and other elements—such as the messages posted by Sarah on her Facebook page—distributed mainly on the net and exchanged informally. An example is offered by the following short conversation:

> ilsegnalatore74: "I heard about a couple of Slavs. But I favor the first hypothesis (voluntary escape). They (the police) should do a house-by-house search in Avetrana. Maybe she is still there, hiding at a friend's."

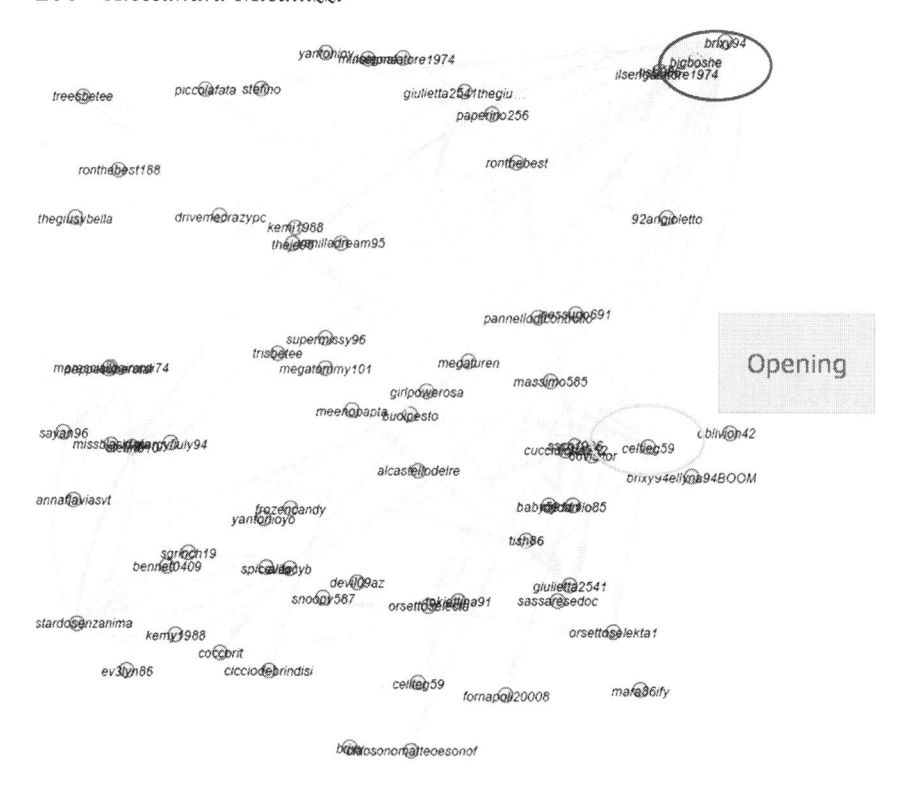

Figure 12.1 Network analysis of the first moment of video participation (red circled: people of Avetrana; orange circled: the contester).

Pannellodicontrollo: "@ilsegnalatore74 no, I think the first hypothesis is wrong. They (friends) said that she had no reason to do that. Moreover, you don't run away from home in broad daylight. It was afternoon!"

During this second phase, the conversational tone is more heterogeneous, and the prevalent trend is to communicate in a distributed manner, without a specific referent, as shown in Figure 12.2.

The map shows how the conversation is distributed among all participants, apart from some cases circled in red, which represent people involved in more than one conversation. Between the lines, it is possible to perceive the *prudery* of knowing more details, the desire of sharing more unknown personal aspects of Sarah. The prevailing theme is the *investigation*, even if there are some isolated cases of flaming toward people who posted comments against the practice of discussing private dramas.

The third phase of participation coincides with the discovery of the victim's body and the public confession by Sarah's uncle. The networked public identified the murderer before the police. The net turned into a public

Anger, Pain, Shame, and Cyber-Voyeurism 201

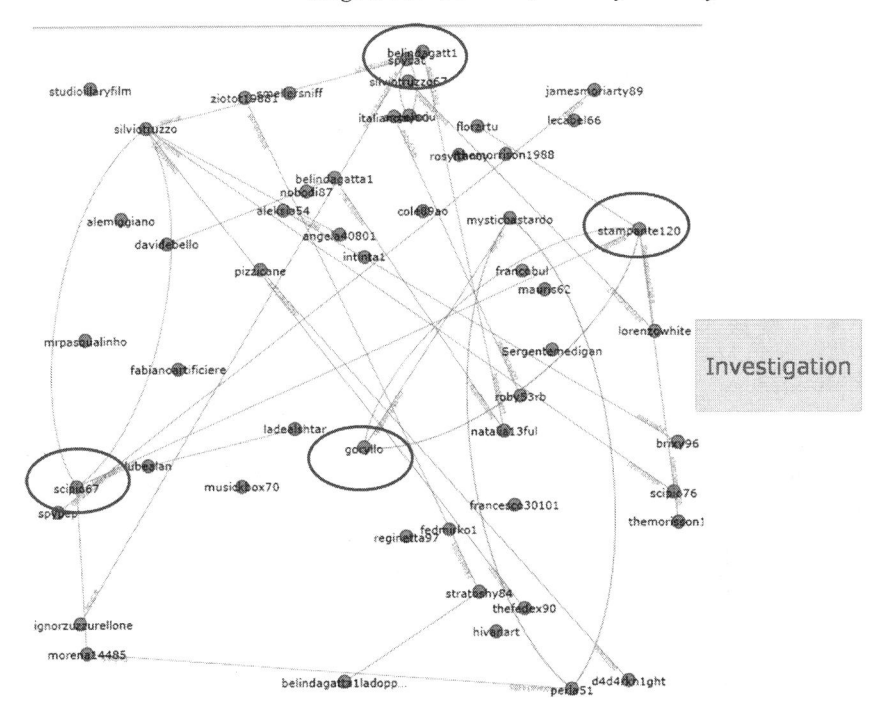

Figure 12.2 The second step of video participation called "investigation" (in red circle: the main hubs of conversations).

tribunal, where everyone expressed anger, hate, disgust, suffering, and grief about Sarah's tragic end. Two other emotions emerging from the messages were particularly evident: the desire for revenge against Misseri (the uncle) and what people identified in him (the destroyer of a happy family image, a pedophile, a murderer, etc.) and shame. A small excerpt of conversations gives an idea of the feelings expressed:

> Nessuno651: "This makes me incredibly furious! It's not fair! She couldn't defend herself! How can he (the murderer) live with this huge burden on his conscience?"

> Nobodi87: "@nessuno651 It was her uncle! I blame him! I can! He was infamous. She trusted him. He saw Sarah grow up. You are a Beast (the uncle), you make me sick!"

> Pietromagno: "I didn't know Sarah, but all this time I have been reading all the news about her. Poor girl! I want to take a train to Avetrana and kick Misseri (the uncle) and his family! Shame on you Misseri!"

Germanotrofio: "He (the uncle) must be made to suffer the tortures of Hell! He should be killed very slowly. I want to do it myself! Monster! Beast!"

This was the most emotive phase of the macro-narration, both in terms of the power and the wealth of nuances of the feelings expressed. The dominant referent was obviously the uncle, to whom all the messages posted were addressed. The posts contain mainly accusations against him, the victim's family, institutions, and some net users who appealed for compassion toward Misseri. But, paradoxically, there were also some messages against the over-exposure of the case by the media.

Interspersed in this complex discussion on both moral and ethical aspects, which stands at the core of the analysis, were also some messages of people who used the circumstance to voice their personal grief, especially about violent or sudden losses. These cases underscore the power of the net as a context where one can experience emotive projection and self-disclosure with a therapeutic effect (Pennebacker 2004; Micalizzi 2010), as exemplified by the following fragment:

Mimmo713: "I know what it means when someone you love dies violently. My cousin was killed in a car accident. He too was an innocent victim. His fault was to cross the street while a drunken driver was passing with his car. Bastard! He is free now and walks around in the streets of my city. We have stopped living ever since that day. I miss him so much."

In the final phase of the analyzed period the tone became softer, and the attention was directed toward Sarah, who was the referent of the messages. It was the time of memory. As shown in Figure 12.3, there was no real conversation, the prevailing trend being to leave a personal message for Sarah, like in a monologist dialogue. The messages were mostly emotive and personal, focused on the grief, suffering, and anger caused by her death.

When looking at the entire sample, it is possible to distinguish four main referents of the messages: Brixy94—the video's author, especially in the first phase; the two alleged murderers (Misseri and later also his daughter Sabrina); and Sarah. Less frequently represented are the media, accused of giving obsessive attention to Sarah's case, and some individual net users. More specifically, the quantitative content analysis shows that the most cited explicit referent of the posts was Sarah, followed by Misseri, Sabrina, and finally by Brixy94, as shown in Figure 12.4.[12]

Regarding the contents of the posts, it was possible to identify six different main themes: accusations, remembrance, emotive support, personal emotions, projection, and investigation. Beside these, there were also some collateral themes, linked with Sarah's case without directly focusing on it: the eternal debate about sexual equality, the conflict between the North

Anger, Pain, Shame, and Cyber-Voyeurism   203

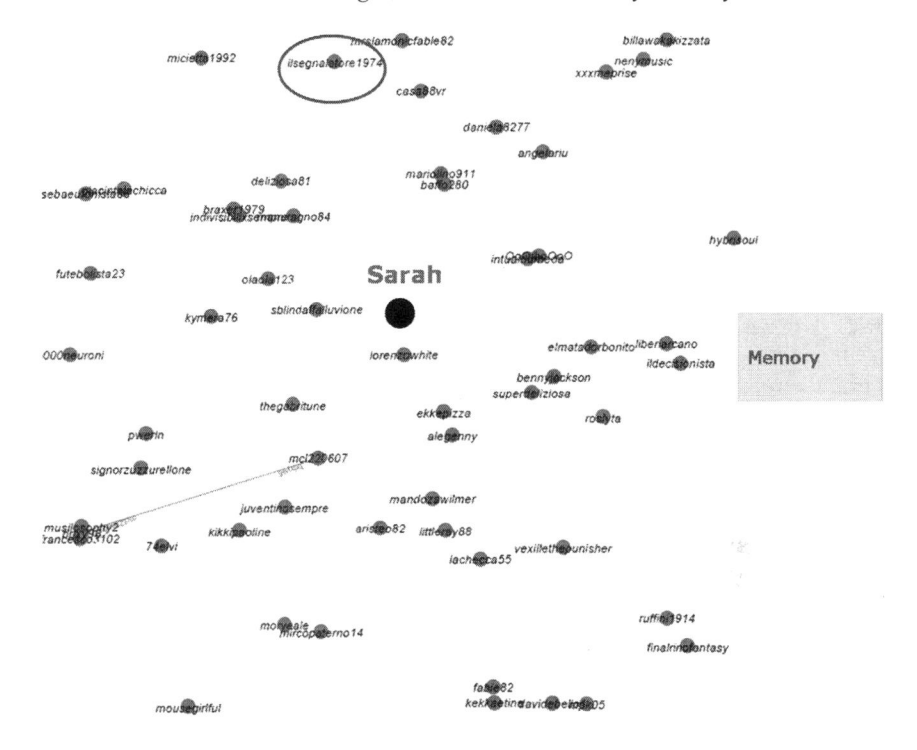

Figure 12.3   The moment of memory in video participation.

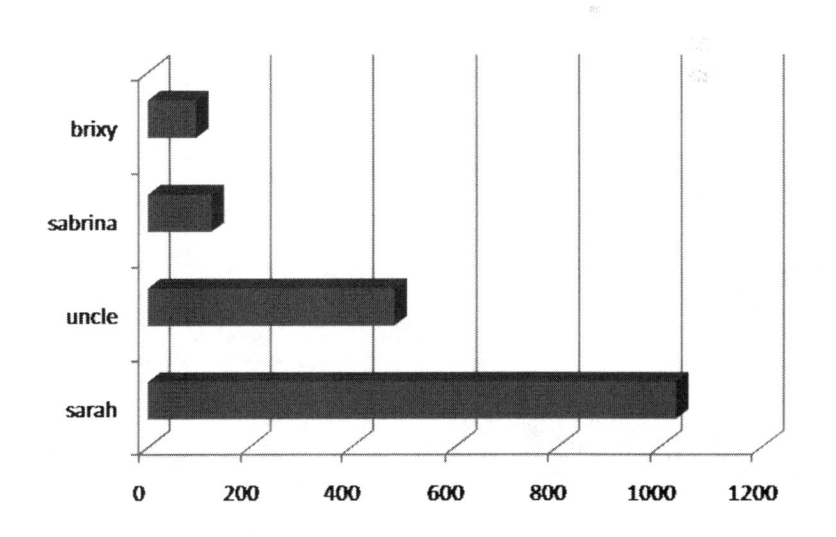

Figure 12.4   Explicit referents of comments.

204 *Alessandra Micallizzi*

text—i.e., all messages exchanged during the period of observation—provided additional findings on the aspect of emotions, specifying which emotions were more narrated than others. Figure 12.4 shows the frequency of occurrence of words semantically related to the emotions that emerged and are specified in the legend. The content analysis was conducted on all 3,117 comments and according to 450 keywords chosen.

The emotion most frequently represented is respect, in the sense of a claim to have pity on Sarah and safeguard her memory. The list of emotions also features "remembrance." Even if it is not considered a feeling, there are two reasons to include this notion—in this specific case—in the sentiment analysis. First of all, emotions play an important role in the memory process (Parrott and Spackman 2000). Secondly, according to the cited authors (Parrott and Spackman 2000) in the first section of this chapter, the act of commemorating a person or an event is characterized by an emotionally intense mood. In other words, publishing a post in the digital space was, in a way, participating in a choral process of constructing memory, expressing personal emotions and coping with them.

As shown in Figure 12.5, the macro-narration was quite rich in emotions. However, if single feelings are combined according to semantic proximity, it is possible to distinguish four groups of emotions. The largest is the group of negative and active emotions. It is the family of revenge, which includes feelings such as anger,[13] disgust,[14] hate, and shame. The use of shame in this specific case is worth noting. We tend to consider shame as a sentiment generated by an individual's exposure to others in a circumstance that breaks social and moral rules (Lewis 2000; Pandolfi

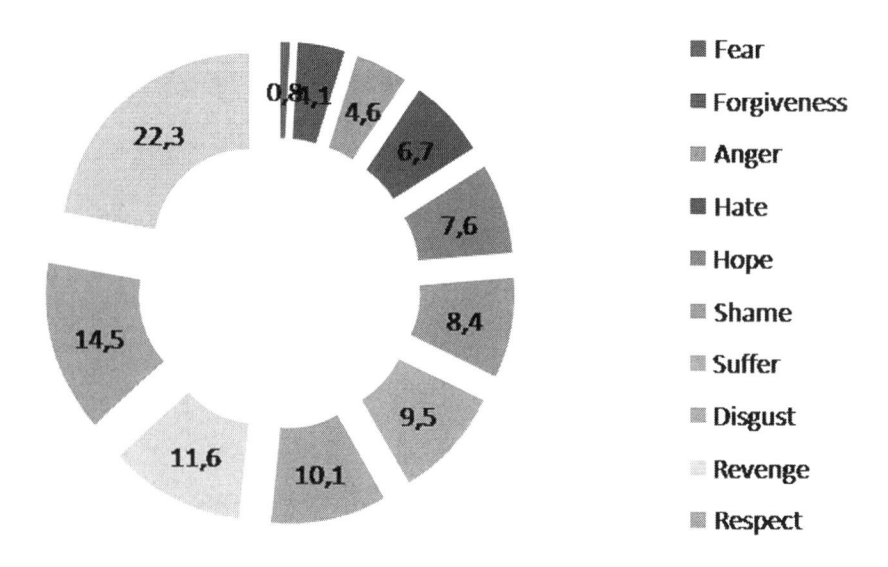

Figure 12.5 Emotion frequencies coming out quantitative content analysis (%).

2002). But in the present context, the word is not used in a reflexive sense. "Shame!" (meaning "you should be ashamed of yourself," or "shame on you!") is used as a kind of war cry, to reinforce the sentiment of disgust and anger about Sarah's death. This usage coincides with the practice of *naming and shaming*, i.e., the disclosure of "information about an identified person or body, which either seeks to induce shame in the person, or at least expresses a judgment that the person ought to feel ashamed of him/herself" (Rowbottom 2012). Hence, this semantic area, which is frequently represented in the conversation, and is exemplified next, may be included in the "revenge" group.

> CicerchiaSeverino: "Misseri, you are a monster. You disgust me. Monster! Jackal! Shame on you. Sarah trusted you, and you betrayed her in this horrible way."

The second group of emotions refers to the grief about the tragic end of Sarah. This group includes suffering, respect, and remembrance (as described earlier).

> Lorenzowhites: "Sarah deserves to RIP! Please, show respect for her. Don't say anything if you don't have anything good to say."

> Theorist2: "Sarah ... she is the girl who ate lemons after school, who wanted to be a waitress, who wanted to be adopted (like Candy, Candy). Do you understand who she was? We will never forget her. She is in our memory. She is a legend!"

> Darek17071: "Being still alive in hearts you leave behind you is not dying; beautiful angel, forgive us and let your ray of light guide us through the darkness, amen."

Less often expressed—and limited to the first phase of the macro-narration—is the semantic area of hope and forgiveness. These emotions, which have a positive and active valence, are less representative of the discussion, having only marginal space and weight.

> Serenopiovoso: "@brixy94 Hey, I hope everything will turn out ok! Don't lose hope!"

> Nessuno691: "He was just a poor man. He is insane, please stop calling for the death penalty and start thinking about forgiveness."

The last group of emotions comprises only one item: the primary emotion of fear.[15] It is generally connected with the process of projection into Sarah's story, mainly by other teenagers who, following this shocking case, began to consider family violence as something likely and near. As is well-known, fear activates a reaction: in this specific case, a narrative reaction.

> Mammana97: "Unfortunately, the truth is that we think we know our relatives and our friends . . . but we only know their good sides. . . . Now I'm scared of men. I'm a teenager. . . . It could happen to me too. . . . I could be in Sarah's place."

From this perspective, the net may be considered as a substitutive social context in which one can express and nurture primary emotions and, more generally, complex feelings (such as revenge and disdain, but also respect and memory) that cause cognitive rumination (Rimé 2009). It seems as if the net becomes the context for cathartic elaboration of strong (and mostly negative) feelings, where physical expression is supplanted by the narrative.

CONCLUSIONS: TOWARD CYBER-VOYEURISM

This chapter examines the practice of social sharing of emotions, introducing the theme through a theoretical framework and focusing on a specific case study of the spasmodic interest of part of the networked public in private tragedies. The study centers on the Narrative-Network analysis (NNa) of over 3,000 posts published as comments to an amateur video on YouTube about an Italian teenager who disappeared in 2010. As evidenced by the study, this type of involvement in so-called private media events (Dayan and Katz 1992) has something in common with the canonic definition of non-pathological voyeurism.

Following the evolution of a tragic event through the net can be viewed as a protected and intimate situation, which the voyeur experiences with no risk to his/her social identity. In other words, it is an *out-of-responsibility* circumstance, and this is confirmed by the reactions of the interviewees. Out of more than 50 successful contacts, only eight people responded to the email proposing an interview, and some of them openly declared their negative reaction to the invitation to "expose" themselves. This indicates that they wanted to watch—to follow the development of the event—without being watched, like a classic voyeur. Moreover, the hostile contents expressed and the huge space for unspeakable thoughts can be considered as part of an obscene act. As some interviewees acknowledged, the prototype cyber-voyeur is interested in all of the tragic private stories told by media news. This is the reason why participation was drastically reduced in the months following the resolution of the case. In fact, attention was transferred to other tragedies, and what is more, there was nothing left to watch.

The practice of cyber-voyeurism has specific characteristics. It is a participated observation, consciously shared with other net users. At the same time, it is public-shed (public and published): the cyber-voyeur leaves traces of his presence by writing messages and exposing them to the public eye. This particular aspect of digital participation can be referred to as the "public keyhole syndrome." In the act of writing messages, the voyeur often makes specific reference to other media contents about the same case and s/he uses the fictional lexicon learned from TV programs. In fact, this practice is semantically contaminated by other media. We can say it is trans-mediated. Finally, the act of reading, writing, and sharing information on the net is also proactive and manipulative, because the voyeur contributes to the evolution of the online macro-narrative of the event. And this is a specific aspect of cyber-voyeurism, since it is possible only through interactive media.

The last aspect deals with the emotional dimension: cyber-voyeurism seems to be characterized by schizophrenic aspects. As mentioned earlier, the voyeur tends to express sentiments of disapproval of and revenge against others, generally in a very hostile manner, but feels no shame for his/her own words. In fact, as theorists of shame specified (Pandolfi 2002), this emotion is associated with exposing oneself in front of relevant others. On the net, the use of nicknames protects the author from the potential judging glances of the Internet audience (Turkle 2011).

In conclusion, the study shows how the digital network replaces physical neighborhood gathering places as a social "chat" context, where people discuss emotions caused by "surprising" and "shocking" events that generate cognitive rumination (Rimé 2009) about emotive mediated quasi-interactions (Thompson 1995).

NOTES

1. One example is offered by the recent Arab Spring, which began with the social sharing of negative emotions on the net. In reading more about these events, it becomes evident that this social effect is complementary, in characteristics and dynamics, to the phenomenon described in this chapter. The Arab Spring is based on social actions generated around a social sentiment, while cyber-voyeurism deals with private cases, particularly emotion-laden ones, that generate social interest and a narrative reaction.
2. Riva, G. (2010) Using Virtual Immersion Therapeutically. In K. Anthony, D. Merz Nagel, and S. Goss (Eds.), *The Use of Technology in Mental Health: Applications, Ethics and Practice* (pp. 114–123). Springfield, IL: C.C. Thomas Publisher.
3. The paragraph is the translation of the definition of voyeurism posted on Wikipedia Italia, retrieved from http://it.wikipedia.org/wiki/Voyeurismo (23/07/2013)
4. As a matter of fact, Sarah's cousin and uncle are both accused of the murder. The trial is still underway as of this writing, and the cousin has been in custody since the end of 2010.
5. This expression paraphrases the more common French term récit-de-vie, used to refer to a fragment of biographical narration, starting from a specific point in the subject's life. In this case, the expression fragment-de-vie is intended to stress the narrative perspective that frames the study and the close correlation between these kinds of posts—highly personal ones—and the identity the net user wishes to project online.

6. For a more specific discussion on the elements of narration, I suggest: Bruner, J. (2003). *Making Stories: Law, Literature, Life*. Cambridge, MA: Harvard University Press.
7. The software applied is T-lab (more information is available on http://www.tlab.it/it/faq.php). Before using a full text content analysis software, the free version of Senti-strength, a specific software for sentiment analysis, was applied.
8. I use the term "macro-narration" to identify all the narrative material analyzed, consisting of all the posts published on YouTube: videos and comments.
9. The software used was Gephy 07.beta.
10. As of September 16, 2010, there were 3,410 comments and 422,216 viewings. The difference between these numbers and those detected during the period of the study is insignificant.
11. He also published a "counter-video" about Sarah, which depicted her negatively.
12. The graph shows the number of times Sarah, Misseri, Sabrina, and Brixy are cited, taking into account all the nicknames and appellatives used to identify each of them.
13. More information about the process of socializing anger is available in: Lemerise, E.A. and Dodge, K. A. (2000) The Development of Anger and Hostile Interactions. In M. Lewis and J.M. Haviland-Jones (Eds.), *Handbook of Emotions* (pp. 594–606). New York and London: Guilford Press.
14. I refer to *moral disgust*, which is a component of the basic emotion of disgust. In fact, disgust "plays a special role in the moral domain as a means of socialization. Insofar as entities viewed as immoral are also disgusting" (Rozin et al. 2000: 644).
15. The surfers use the term "fear," but the emotion expressed is actually anxiety. As described by Oham (2000), these two feelings are strictly connected and can be activated by the same emotional stimulus.Notes
16. rumination (Rimé 2009) about emotive mediated quasi-interactions (Thompson 1995).

REFERENCES

Bakardjieva, M. (2003) Virtual Togetherness: An Everyday Life Perspective. *Media, Culture & Society,* 25, 291–313.
Boccia Artieri, G. (2010) Conversazioni non mediate sulla crudeltà. *Apogeonline*. Retrieved from http://www.apogeonline.com/webzine/2010/09/08/conversazioni-non-mediate-sulla-crudelta (23/07/2013)
Bolter, J.D. and Grusin, R. (2002) *Remediation: Understanding New Media*. Milan: Guerini e Associati editore.
boyd, d. (2007) Why Youth (Heart) Social Network Sites: The Role of Networked Publics in Teenage Social Life. In D. Buckingham (Ed.), *MacArthur Foundation Series on Digital Learning, Youth, Identity and Digital Media Volume* (pp. 119–142). Cambridge, MA: MIT Press.
Dayan, D. and Katz, E. (1992) *Media Events: The Live Broadcasting of History*. Cambridge, MA: Harvard University Press.
Flichy, P. (2007) *The Internet Immaginaire*. Cambridge, MA: MIT Press.
Gump, B.B. and Kulik, J.A. (1997) Stress, Affiliation, and Emotional Contagion. *Journal* of *Personality and Social Psychology,* 72, 305–319.

Anger, Pain, Shame, and Cyber-Voyeurism 209

Kazys, V. (Ed.). (2008) *Networked Publics*. Cambridge, MA: MIT Press.
Kemper, T.D. (2000) Social Models in the Explanation of Emotions. In M. Lewis and J.M. Haviland-Jones (Eds.), *Handbook of Emotions* (pp. 45–74). New York and London: Guilford Press.
Lemerise, E.A. and Dodge, K.A. (2000) The Development of Anger and Hostile Interactions. In M. Lewis and J.M. Haviland-Jones (Eds.), *Handbook of Emotions* (pp. 594–606). New York and London: Guilford Press.
Lewis, M. (2000) Self-Conscious Emotions: Embarrassment, Pride, Shame and Guilt. In M. Lewis and J.M. Haviland-Jones (Eds.), *Handbook of Emotions* (pp. 623–636). New York and London: Guilford Press.
Luminet, O., Bouts, P., Delie, F., Manstead A. and Rimé, B. (2000) Social Sharing of Emotion Following Exposure to a Negatively Valenced Situation. *Cognition and Emotion*, 14(5), 661–688.
Micalizzi, A. (2010, May) Mourning Network: Ethnography and New Social Practices in Online Communities. *Digithum*, I–VI.
Micalizzi, A. (2011, September) *Narrative Network Analysis (NNa): Un approccio integrato per lo studio della socializzazione in Rete*. Paper presented at the RiQ-GioS International Conference on Qualitative Methods, Kore University, Italy.
Oham, A. (2000) Fear and Anxiety: Evolutionary, Cognitive and Clinical Perspectives. In M. Lewis and J.M. Haviland-Jones (Eds.), *Handbook of Emotions* (pp. 573–593). New York and London: Guilford Press.
Ong, G. (1982) *Orality and Literacy*. New York: Routledge.
Pandolfi, A.M. (2002) *La vergogna: Un affetto psichico che sta scomparendo?* Milan: Franco Angeli.
Parrott, W.G. and Spackman, M. (2000) Emotion and Memory. In M. Lewis and J. Haviland-Jones (Eds.), *Handbook of Emotions* (2nd ed.; pp. 476–490). New York: Guilford.
Pata, K. (2011) Participatory Design Experiment: Storytelling Swarm in Hybrid Narrative Ecosystem. In B.K. Daniel (Ed.), *Handbook of Research on Methods and Techniques for Studying Virtual Communities: Paradigms and Phenomena* (pp. 482–508). Hershey, PA: Information Science Reference.
Pennebacker, J.W. (1997) Writing About Emotional Experience as a Therapeutic Process. *Psychological Science*, 8, 162–166.
Pennebacker, J.W. (2004) *Writing to Heal: A Guided Journal For Recovering From Trauma and Emotional Upheaval*. Oakland, CA: New Harbinger Press.
Peters, J.D. (2011) Witnessing. In P. Frosh and A. Pinchevski (Eds.), *Media Witnessing: Testimony in the Age of Mass Media* (pp. 23–41). New York: Pallgrave Macmillan.
Rimé, B. (1995) The Social Sharing of Emotional Experience as a Source For the Social Knowledge of Emotion. In J.A. Russell, J.M. Fernandez-Dols, A.S.R. Manstead, and J.C. Wellenkamp (Eds.), *Everyday Conceptions of Emotions: An Introduction to the Psychology, Anthropology, and Linguistics of Emotion* (pp. 475–489). Dordrecht: Kluwer.
Rimé, B. (2009) *Le partage social des émotions*. Paris: Presses Universitaires de France.
Riva, G. (2010) Using Virtual Immersion Therapeutically. In K. Anthony, D. Merz Nagel, and S. Goss (Eds.), *The Use of Technology in Mental Health: Applications, Ethics and Practice* (pp. 114–123). Springfield, IL: C.C. Thomas Publisher.
Rowbottom, J. (2012, February) To Punish, Inform and Criticize: The Goals of Naming and Shaming. *Proceedings of the International Conference on Media and the Boundaries of Disclosure: Media, Morals, Public Shaming and Privacy*. Oxford, UK.

Rozin, P., Haidt, J. and McCauley, C. (2000) Disgust. In M. Lewis and J.M. Haviland-Jones (Eds.), *Handbook of Emotions* (pp. 637–654). London: Guilford Press.

Shott, S. (1978) Emotion and Social Life: A Symbolic Interactionist Analysis. *American Journal of Sociology*, 84, 1317–1334.

Thompson, J.B. (1995) *The Media and Modernity: A Social Theory of the Media*. Cambridge, MA: Polity Press.

Turkle, S. (2011) *Alone Together: Why We Expect More From Technology and Less From Each Other.* New York: Basic Books.

Walzer, J. (2001, September) *Distributed Narrative: Telling the Stories Across Network.* AoIR Conference, Brighton. Retrieved from http://jilltxt.net/txt/Walker-AoIR-3500words.pdf (23/07/2013)

13 Emotional Investments
Australian Feminist Blogging and Affective Networks

Frances Shaw

Over three years, between 2008 and 2011, I researched a network of feminist bloggers in Australia. I collected data from blogs in the network, used a set of seed links from the main blogs in the network to map the network over time, and interviewed 20 key actors from within the network. These interviews took place all over Australia, and from these interviews I drew out a number of themes that emerged in our conversations. One of the themes was the emotional and affective dimensions of feminist bloggers' relations with one another in the network, and with mainstream media and society. This chapter addresses this theme from my research, and draws out a number of ways in which emotion and affect has not only personal significance, but also political significance to participants in the network. It argues that the building of community, and the negotiation of feminist claims out of an affective reaction to the mainstream, have a political purpose and outcome.

In the conversations that I had with them, Australian feminist bloggers explained a number of ways that affect and affective relations have impacted on their involvement in the blogging network. Bloggers are shown to develop an affective relation toward discourses and events in the media and in mainstream politics and society. In my research I found that emotional investment and affective relations are part of the practice of counter-hegemonic politics. This chapter describes that relationship and how it is constructed in the context of a network of feminist bloggers.

The Australian feminist blogging community is a discursive 'safe space' in which feminist claims are generated. Firstly, bloggers collectively develop an affective relation toward discourses and events in the media and in mainstream politics and society. Women explained that they began blogging and continue to blog because it is a way to push back against things that they disagree with in the mainstream. This is shown in the recurring theme of the blog as 'outlet' when discussing their motivations to blog. It is also shown in the way that their responses to media events are framed in terms of their own affective responses to those media events and discourses. Affective motivations for political response, far from leading us to the conclusion that feminists engage in an illegitimate form of emotional politics, as some critical comments on feminist blogs suggest, instead point toward Tomlinson's (2010) work on emotionality in political discourse as a rhetorical device. In this view, emotionality is not only tactical, but at the core of political engagement.

Secondly there are the relations of antagonism and aversion that online feminists hold toward 'trolls' and others in the Internet-based discursive space, as well as parts of the mainstream media. In her work on Internet counter-publics and cyber-movements, Palczewski (2001: 172) argues that safe spaces are essential for the development of counter-hegemonic claims. She defines safe spaces as "a space where exploratory discourse is possible, where one is able to make mistakes knowing the opportunity to correct them exists" (Palczewski 2001: 172). In what follows, I explore the concept of the 'safe space' and how it is described by feminist bloggers, as well as the ways that their narratives of participation contradict this idea of 'safety.' I also explore the way that feminist bloggers describe the practices and defenses that they have built up to repel trolls and disruptive others. Bound up in this aversive politics are the practices of moderation that feminist bloggers have developed to delimit allowable expressions, a practice of defining the offensive that disallows these discourses from entering the 'safe spaces' of feminist blogs, except in opposition.

Finally, this chapter demonstrates how community has become an important component of Australian feminist blogging practice, and women engage in a number of different kinds of work to ensure that the community builds in size and strength of affinity. Bloggers develop affective attachments toward one another, build friendships, and build and maintain a sense of community. This can range from making an effort to read and link to other Australian blogs, to running, submitting to, curating, and hosting the *Down Under Feminists' Carnival*. It can also involve the efforts of individual bloggers to be inclusive, supportive, and welcoming to others.

BLOGGING AS OUTLET

Feminist bloggers use blogs to respond negatively to mainstream discourse, as a result of the disjuncture between their own political views and the mainstream. They express these responses through blogs in a politically engaged way, and connect with others who experience similar reactions. There are emotional advantages to the practice of blogging among like-minded others, but the purpose of my argument here is to show the ways that this is also political.

In feminist blogs, there is a shared idea about the expression of anger as a good, in political terms, and as both permissible and positive. The injunction to embrace anger is present in the title of the blog *Fuck Politeness*. The blogger 'Fuck Politeness' (personal communication, December 15, 2009) explains how she had seen in the media instances of when "politeness was used to shut down debate" and would override "somebody's life and what was going on," and decided that *Fuck Politeness* would be a good name for a blog. This links strongly to Tomlinson's (2010: 60) claim that conventions of civility "serve entrenched interests by encouraging aggrieved parties to give up part of their bargaining power—their emotional force and its consequences—*prior* to negotiation" (emphasis hers).

'Fuck Politeness' decided to start writing her blog as a result of the frustrations she felt in everyday conversations and arguments with others, because she felt that her ideas were not listened to, and felt that writing might be a way to actually get her ideas out there. She felt blocked from contributing to political conversations in her everyday life because of the way her style of communication was discredited, and found, in blogs, a way to express her anger that was accepted and even encouraged. Her blogging was a response, not just to problematic discourses, but also to her exclusion from those conversations.

> I remember that being something that really got me fired up about writing and about having a space where my voice could actually be listened to because I was so used to that being the end result of any disagreement with a guy where [frustrated laugh] they could just do that, they could just say "Oh, you're being oversensitive," "Oh, you're being stupid," or "Oh, you're just a man-hater." And I think it really hit me how little space there was for me to actually push back and have the last word and say, "You're not listening to what I'm actually saying." I don't know, I guess I got sick of just ranting about these things to my friends and thought that I might as well have a go at writing. ('Fuck Politeness,' personal communication, December 15, 2009)

For so many Australian feminist bloggers, blogging is often motivated by a desire to let out anger. "I think I write better when I'm angry," one woman told me ('Caitlinate,' personal communication, March 8, 2010). More than one participant told me that they write in blogs so as not to rant at friends and family members, to put their political feeling into something productive and positive:

> I think it's really nice to have an outlet for that rage. One of the great things about having a blog and to be able to write in your own way, is that you don't have to bore your own partner and your close friends with the same things. ('CrazyBrave,' personal communication, March 10, 2010)

And again, from 'Spilt Milk':

> I guess that's an example of me seeing something in the media and just being very cross, [Laughs] and in the past I might have just said

something to my husband or a friend if I was talking to them on the phone. You know, "did you see that? It's terrible," and then that would've been it. But now that I have a blog I can actually rant at people. I think that's an example of one of the really positive things that come out of blogging for me I think is that it's that kind of outlet so if I do feel outraged by something I can express that and I hope that some people read it and concur or some of the people I know in real life might read it and go "oh I never really thought about that." ('Spilt Milk,' personal communication, January 30, 2010)

'Tigtog' explained that "blogging is very much about your own personal levels of outrage or disgust or disdain," and thus writing politically through blogs is not necessarily a systematic process—"You just can't concentrate on everything" ('Tigtog,' personal communication, December 12, 2009). Writing blogs is bound up in enjoyment and the push to create and share ideas with people who will want to hear them, and who will appreciate and identify with the blogger's perspective. The subject matter that a blogger chooses to write about is motivated by an affective response to an event or experience or problem, and a desire to respond to it:

> I find what's more important, what's more useful to me is a strong desire to speak rather than having any sort of idea of the shape or the tone or the mode of the finished product. [. . .] It's more about the impulse. ('Lucy Tartan,' personal communication, March 8, 2010)

Blogs are an outlet for not just anger, but also creativity, enjoyment, fun and humour, analysis, sharing, and distributions of attention—for example through link posts and 'signal boosts'—and anything that is not fulfilled in other areas of the blogger's life. This can be because of childcare or work commitments, or in response to the limitations in the kind of conversations that are possible with one's immediate peer group or closest family. One woman told me that blogging was "a real outlet for me because I get very bored, being a stay at home mother. I need more intellectual stimulation than my two year old can give me, [. . .] and also it's a bit of an emotional outlet sometimes too" ('Spilt Milk,' personal communication, January 30, 2010).

Not everyone that I interviewed felt that blogging necessarily provided them with an emotional outlet, but reframed the idea of 'outlet' in other terms: "I tend not to get angry at things," one blogger explained, "I get analytic about them" ('In A Strange Land,' personal communication, March 4, 2010). Sometimes it is also about wishing to engage in the political and putting their political opinions out there in the community, about contributing to political discussion:

> Definitely sometimes it's an emotional impetus. There are some of my posts which are very much driven by being pissed off at something or

thinking that something is really irritating or so forth. But there are also quite a few that are more reflective, like they're more based on a sort of analytical response, so I think it's a mixture of both. ('Godard's Letterboxes,' personal communication, March 9, 2010)

This idea of blogging as either 'analytical outlet' or emotional response hints at the link between the affective and the individual as political subject. Bloggers respond to dissonance in their sense-making structures both intellectually and as a response to political feeling. These responses cannot be separated. Both occur as a result of something jarring in mainstream discourse that then leads to a discursive response. Bloggers develop oppositional discourses around particular mainstream discourses, and this leads to conflict with that outside discourse. But there are also conflicts within blogging communities, between bloggers in the negotiation of feminist politics, and between bloggers and outsiders who disrupt and harass them.

TROLLS AND OTHER DISSONANCE

For women who use their blogs as an outlet for personal frustrations and political feeling, as well as emotional support from others, it can be a jarring experience when other people come to a blog in order to troll, to intentionally bait bloggers and derail discussions. This is especially true when people use blogs as a personal outlet, which can give the community the atmosphere of a support network. If bloggers often speak of their intent to create a safe space for feminist discussions, this is in direct tension with another subset of visitors to feminist blogs—those whose views are directly opposed to feminism.

Feminist bloggers have devised a number of strategies to deal with people who come to feminist blogs to disrupt or cajole. A theme that came up again and again in my interviews with feminist bloggers is the problem of trolls and trolling, moderation policies, and issues surrounding those scenarios and practices.

Because of the culture of strong moderation policies in place in the network, troll attacks are now rarely visible, because they are either deterred, moderated, disemvowelled[1], deleted, or ignored. Trolling is often only visible as a note in the comments or the post itself. In their strategies to deter trolls, feminist bloggers occasionally 'call troll' and discuss the trolling event, to make visible particular behaviors and tactics that trolls use to derail discussion.

Such discussion also makes visible the aggressive attention that women—particularly feminist—bloggers receive. One example of this is *Hoyden About Town's* 'Troll-Off!' ('Lauredhel,' February 17, 2009), a satirical poll where participants could vote for one of seven trolls. Trolling comments, though moderated or deleted in the original entries, were collected and reproduced in the poll article and posted together. In this way they were clearly tagged as unacceptable. Tomlinson (2010: 145) writes on the rhetorical strategy of 'intensification,' wherein scholars "quote, echo, aggregate, exaggerate, and in other ways appropriate the language of other people's pedagogical texts." The satirical aggregation of examples of trolling has the same effect, 'heaping' and 'conglomerating' language to demonstrate its superfluity (Tomlinson 2010: 145).

The existence and expression of these opinions therefore make antifeminist viewpoints more visible, and radicalize feminists who had previously assumed that these opinions were not widely held. Once engaged in online discourse, the feminist blogging community can serve as a safe space away from such discourses and opinions, as well as a space to talk about and counter those views. As 'Ariane' (personal communication, November 24, 2009) explains, being part of the feminist blogging community enables her to engage with political ideas without having to interact with "people whose opinions I really can't handle":

> One of the things I like about it is that I can see all these different perspectives and opinions that are still reasonably well aligned with my own, so that it's not just depressing, and I get a sort of filtered view of the outside world by people reacting to it. ('Ariane,' personal communication, November 24, 2009)

'Ariane' feels there are people in the mainstream with horrendous ideas and opinions, and in feminist blogs she finds protective space away from them where she can engage with like-minded people. 'Rayedish' expressed a similar view:

> I've found that I've definitely become more feminist in who I want to read and why because I found it's sharpened my sensitivity to misogyny in comments and stuff, and I just don't want to read it if I don't have to. A feminist blog [is a] safe space. ('Rayedish,' personal communication, March 17, 2010)

Feminist bloggers have developed collective strategies to counter trolling and harassment when it occurs. They have developed these strategies in response to the aggression, harassment, and kneejerk responses that they face in comments. These strategies are not only practical, but help to build relationships within the community. Building up an awareness of how to deter and discourage trolling helps to build camaraderie and a sense of safety in the community. 'Godard's Letterboxes' told me how it helps her to know that she's part of a group dealing with the same issues and problems:

Particularly if you write something on rape and get trolling comments back at you, you don't feel alone in that you're the only one who is facing those kinds of things. And that you can have the courage to just say, "Your opinion doesn't really count" or "I'm not interested in you" and looking at how other people deal with comments and things is quite interesting and quite empowering. ('Godard's Letterboxes,' personal communication, March 9, 2010)

The way that feminist bloggers have developed strong moderation policies in cooperation with each other is closely related to the development of the feminist community itself. In particular, bloggers make use of backchannels such as Twitter, email lists, and personal emails and chats to share solidarity when they come up against particular forms of opposition, such as trolls, or extremely problematic political statements or attacks.

Feminist bloggers see the shared moderation culture as a way to be supportive of one another, and to prevent others from having to deal with offensive attacks or 'triggering' ideas. 'Blue Milk' ('Blue Milk,' personal communication, February 24, 2010) explained that one of the reasons she avoids reading big mainstream political blogs is that they are not as well-moderated as the big feminist blogs who "look out for that sort of trolling behaviour." She values sites like *Hoyden About Town* because "you know you're not going to get mercilessly attacked by a bunch of trolls" ('Blue Milk,' personal communication, February 24, 2010). Moderation is a responsibility and a labor that feminist bloggers commit to when deciding which comments to publish and which to not publish. Moderation has negotiated conventions, individually enforced, and bloggers make personal decisions about where to draw the line. The feminist community has developed a culture of moderation:

> I think people are a lot more comfortable now telling people that they're being off topic. There's a lot more understanding of the different styles of trolling that are used to disrupt a discussion and derail it off onto something inconsequential. ('Tigtog,' personal communication, December 12, 2009)

Long-time bloggers such as 'Tigtog' derive their tactics toward trolling commenters from their experience of trolling in their use of the internet going back to Usenet newsgroups and forums. 'Tigtog' (personal communication, December 12, 2009) argued that her experience in such communities helped her to have a realistic understanding of internet debate, derailment, and trolling practices, enabling her to develop a strong moderation culture in the *Hoyden About Town* blog particularly. Such moderation practices promote a sense of safety and community that aims to allow feminist discourse to flourish.

THE POLITICS OF FEMINIST SPACES

Bloggers describe the development of designated spaces for feminist thinking and writing as a political act. Among other things, the blogging network functions as a support network for feminist and politically active women. "A lot of people have been pretty badly damaged by other people being horrific to them," Ariane explains, and because of this is sometimes "feels more like a support network than a movement for change" ('Ariane,' personal communication, November 24, 2009). But building an affective community was also often understood as part of the political work of blogs. Bloggers see engagement and interaction within the feminist community as part of their feminist practice.

Internet communities also, ideally, provide a greater degree of accessibility and inclusion. Feminist bloggers strive to make this a reality, and some of the women that I interviewed explained that blogging allowed them to engage politically in ways that they otherwise would not be able to. Blogs allow people to find like-minded people in geographically dispersed locations. The community also allows for international engagement with bloggers overseas. The international feminist community was not within the scope of my study. However, over the three years of this research, Australian feminist bloggers forged strong links and maintained relationships with a range of bloggers in other locations—the United States and New Zealand, in particular. As 'Caitlinate' (personal communication, March 8, 2010) explains, the Internet-based feminist community can provide replacements "for real life communities for people who are alienated or physically unable to access a wider physical community whether through their own abilities or through being in the middle of nowhere or something." For women who are raising children, or whose access to spaces outside their homes is otherwise limited due to disability, and/or who live in places where their contact with other feminists is limited, the blogging network is invaluable:

> [The] online community is quite important to me, because I don't have a huge number of people in my real life connections with similar outlooks on life. ('Mimbles,' personal communication, February 23, 2010)

Backchannels such as Twitter and to an increasing extent Tumblr have also developed an importance in reinforcing and encouraging a sense of community among feminist bloggers:

> Before, even when I was reading blogs every day, I still felt sort of apart from it because I wasn't commenting on it. Now I'm talking to those same people on Twitter, and even when I'm not commenting on their blogs, I'm still talking to them about stuff, so it's actually a really powerful social tool. ('PharaohKatt,' personal communication, February 28, 2010)

Women engage in a number of different kinds of labor to build Australian feminist blogging communities. This can range from making an effort to read and link to other Australian blogs, to running, submitting to, curating, and hosting the *Down Under Feminists' Carnival*. Caitlinate explains why she thinks the development of community is so important politically. Her response highlights the emotional dimensions of her political involvement:

> Feeling that community power, and that spoken power and that written power, is really really really amazing and much more powerful than just being someone explaining really clearly why that's not okay, I think. Even if it's not said explicitly, the emotion of it and the fact that these people are gathering together to say that that's not okay and here are all these things that happened and we see them and we know that they're there and as a community we fight them or speak out against them. That's incredible. ('Caitlinate,' personal communication, March 8, 2010)

The blogging community thrives on networks of attention. Tigtog explains how she uses link posts to build a sense of community and help Australian feminists to find one another. This shows how the building of networks of attention can in fact be an intentional and political activity rather than simply being informed by personal interests. Directing others' attention to different voices enables different voices to be heard:

> I think one of the best things of the blogosphere generally is the way that you can lead your own readers to find other voices. You find other voices yourself through what you read on people's blogs, that they find somebody and linking to them, you can sort of help other people find them and build them up and whatever. ('Tigtog,' personal communication, December 12, 2009)

I would argue that while many of the early blogging participants came to the Australian blogs through contacts in North American blogging networks, the current strength of the Australian blogosphere and in particular the present-day Australian feminist blogosphere really had to be constructed through the links and attention that Australian feminist bloggers gave to their own community. This is something that takes a great deal of effort and is a political labor. An example of this kind of work is the aforementioned *Down Under Feminist Carnival*, which has been running for a little over two years, was begun by 'Lauredhel' from the blog Hoyden About Town, and is now being administrated by Chally Kacelnik. This labor is appreciated and understood by other bloggers as important to the health of the community, and as having solidified and extended the networks of attention that existed at its inception.

COMMENTS

Comment discussions have a particular importance in blogs, allowing for clarifications and argument that is associated with spoken conversation (Barlow 2008: 16). Because comments, unlike the original blog posts themselves, cannot generally be edited after posting, they are more like direct speech which "can only be amended through an overlay of other words" (Barlow 2008: 15). Bloggers expressed the fact that developing a culture of debate on their blogs requires a certain level of commitment and work on their own part, in order to develop a sense of a community for dialogue and challenge:

> [M]y blog doesn't have a culture of the commentariat talking to each other in the threads. I'm actually actively trying to encourage it at the moment by replying more often to people in the threads, but they don't tend to talk to each other. ('Mimbles,' personal communication, February 23, 2010)

They give this debate a positive value and work to encourage a culture of debate:

> I make an effort to respond to every comment, you know, so it does become a discussion rather than me saying "this is my opinion!" ('News With Nipples,' personal communication, December 14, 2009)

While debate is viewed positively, it can also be a stressful challenge, particularly if debate is supplemented by abusive or trolling comments, or community conflict. Chally Kacelnik, in discussing the things that have affected her emotionally as part of the community, talked about how stressful moderating comments can be when 'the blog wars' are on:

> [T]here's been blog drama after blog drama, and I've mostly not been in those but the ones I've had, it's been quite distressing. But I try to remember that it's going to happen again and again and I need to just stay focused on doing the valuable work that we need to do and be in community with these people, because we can't all get along but most of the time we can, so that's what I'm trying to do. (Chally Kacelnik, personal communication, November 25, 2009)

As well as by commenting on each others' blogs and creating networks of attention, bloggers also use backchannels such as Twitter and Facebook to reinforce a sense of community and to strengthen and indeed extend the network and influence of the online feminist community. Although I have focused on a blogging network in this study, such backchannels have increasing significance for the reach and spread of the ideas that are generated and reinforced within this community.

Emotional Investments 221

Bloggers consolidate their relationships with one another in other ways apart from reading and commenting on each others' blog posts, and interaction in online backchannels. Many bloggers, though by no means all of them, develop offline relationships with people in the community. They also oftentimes see no difference between this kind of socialization and 'hanging out' in backchannels such as Twitter. As 'Mimbles' (personal communication, February 23, 2010) explains: "I like that you can have ten, fifteen minute conversations with three or four people while just sitting here at home." While Tumblr is another blogging platform, many bloggers use it in addition to their regular blog to post links and images and shorter posts, and it is conducive to building friendships and community off of the main sites where traffic is heaviest. Tigtog discusses the value of backchannel communication:

> There's a great deal of solidarity in particular behind the scenes. I'm on several different mailing lists that are made up of different cross-sections of feminist bloggers who've for one reason or another have put the mailing list together and keep in touch with each other about things, and it's nice sometimes to be able to pose a question to those people and get an answer back. ('Tigtog,' personal communication, December 12, 2009)

Conversations held through Twitter serve to help bloggers share aspects of their personal lives with one another, make small talk, and chat to amuse themselves. It is often less serious talk than the conversation that takes place in comments and in the dialogue of blog posts, but it performs a very important function in the community, cementing personal ties and making linkages visible. It helps people find new people to connect with, and to find new things that they have in common in addition to feminist politics. It also helps to rally people at times of crisis, by providing links to posts or issues that come up in the news and on other blogs.

POLITICAL EMOTIONS AND AFFECTIVE NETWORKS

The bloggers that I spoke to used their blogs to give form and shape to the dissonance they felt in their own lives, and to share the discourses that enabled them to turn it into political claims. Blogging allowed them to focus their anger and think it through. They were able to articulate particular claims in response to mainstream discourses that made them angry or otherwise upset. These articulations provided the justifications for further confrontations with the mainstream, not only for them, but also the others that they have now *armed with* the discourse to do so.

> That is an incredible benefit of taking part in blogs, is feeling, is finding so many other people with views like yours, and not feeling so alone,

and getting to really develop your feminism because I think otherwise it can stall. ('Blue Milk,' personal communication, February 24, 2010)

The affective relation to the mainstream that is constructed within the community, and the simultaneous negotiation of ways to react, I argue, provide a means for the political that was not there before. This is related both to the creation of political subjectivities, and participation in discursive activism. Emotion is frequently political in itself, rather than simply creating the preconditions for political action, because it occurs as a result of the creation of a political subject who would respond. This response is Rancière's 'dissensus,' and the result of 'rupture' (Rancière 2010). Affect, though frequently the result of women's relations to their particular worlds, finds its place and echo in the writings of other women. Women found the articulation of their political feeling to be most powerful when other women could identify with that feeling. The outlet for emotion was frequently a political analysis, but one that included affect as part of its rhetorical push.

Feminist bloggers use a number of tactical strategies to highlight these affective aspects of blogging. The first is the 'trigger warning,' which highlights to readers that particular content may recall particular traumas or affective responses to oppressive discourses. The most common example of this are posts about rape or violence against women, but extends to a number of other potentially inflammatory topics such as dieting talk.

In this chapter I have argued that political affect is an important part of processes of articulation, identification, and activism. I have shown how feeling impacts on the development of this political community. Anger, care, and support are acted out by women not only out of emotional need, but also performed as part of their political practice. A feminist understanding of social movements needs to take affective rhetoric seriously, and not devalue emotional discourse in political debate.

In my research, I found that political emotions have been part of the process of articulation, identification, and indeed activism. Feminist bloggers "use affect in constructing alternative rhetorics" (Tomlinson 2010: 20). Emotion is frequently political in itself, because rather than creating the preconditions for political action, it occurs as a result of the creation of a political subject who would respond. Affect, though frequently the result of women's relations to their particular worlds, finds its place and echo in the writings of other women. Women found the articulation of their political feeling to be most powerful when other women could identify with that feeling. The outlet for emotion was frequently a political analysis, but one that included affect as part of its rhetorical push.

The study of affect should properly "illuminate [. . .] both our power to affect the world around us and our power to be affected by it, along with the relationship between these two powers" (Hardt 2007: ix). The intimate relations that are generated within the feminist online community are part of the process of writing together a feminism (or feminisms) that is responsive to the changing social environment. As feminists in the community have argued, counter-feminist rhetoric is more visible than ever in the words of trolls and vocal minorities in online spaces. The development of a support network for feminists in the feminist blogosphere should not be understood in any way as a withdrawal from the political, except in a sense that it is a withdrawal from those things that they are politically averse to. Instead it is a space in which the development of feminist ideas are safe from attack, through the practices of moderation policies and a careful commitment to discursive practices that are not exclusionary.

In these blogging networks, the personal acted on the political through the negotiation of political discourse. Rather than delinking women's problems from the political, the discourse of the blogging network linked women's experiences to the political. In many ways the community's purpose was as a 'support network' for feminists, as some of my interviewees described it, but simultaneously it provided the material for explicitly political interventions in the mainstream. While the work of discursively linking affect with the political is performed through discursive activism, the status of the community as a support network is a strong binding force in making the community a political one.

Language that draws on affect can be understood as laying claim to a particular moral force. Additionally, affective relationships and networks can bind political communities together in ways belied by their often nonidentical interests and beliefs. Women in the community told me that because of the relationships they have built in the community, and the respect that they have for other participants, they were able to change their own minds and negotiate what feminism should be.

NOTES

1. The practice of removing vowels to make a troll's comment unintelligible.

REFERENCES

'Ariane'. (2009, November) Interview.
Barlow, A. (2008) *Blogging America: The New Public Sphere*. Westport, CT: Praeger.
'Blue Milk'. (2010, February) Interview.
'Caitlinate'. (2010, March) Interview.
'Crazybrave'. (2010, March) Interview.
Down Under Feminists' Carnival. (2011) *Down Under Feminists' Carnival*. Retrieved from http://downunderfeministscarnival.wordpress.com/. Accessed March 2, 2012.
'Fuck Politeness'. (2009, December) Interview.
'Godard's Letterboxes'. (2010, March) Interview.
Hardt, M. (2007) Foreword: What Affects Are Good For. In P. Clough and J.O. Halley (Eds.), *The Affective Turn: Theorizing the Social*. Durham, NC: Duke University Press.

'In A Strange Land'. (2010, March) Interview.
Kacelnik, C. (2009, November) Interview.
'Lauredhel'. (2009, February 17) Troll-Off! Vote Now for Your Favourite Troll. [Blog post]. Retrieved from http://hoydenabouttown.com/20090217.3779/troll-off-vote-now-for-your-favourite-troll/
'Lucy Tartan'. (2010, March) Interview.
'Mimbles'. (2010, February) Interview.
'News With Nipples'. (2009, December) Interview.
Palczewski, C. H. (2001). Cyber-movements, New Social Movements, and Counterpublics. In R. Asen & D. C. Brouwer (Eds.), *Counterpublics and the State*. Albany: State University of New York Press.
'PharaohKatt'. (2010, February) Interview.
Rancière, J. (2010) *Dissensus: On Politics and Aesthetics*. London: Continuum.
'Rayedish'. (2010, March) Interview.
'Spilt Milk'. (2010, January) Interview.
'Tigtog'. (2009, December) Interview.
Tomlinson, B. (2010). *Feminism and Affect at the Scene of Argument: Beyond the Trope of the Angry Feminist*, Philadelphia: Temple University Press.

Contributors

Tova Benski is a senior lecturer at the Department of Behavioral Sciences, the College of Management–Academic Studies, Rishon Lezion, Israel. Her fields of academic interest and research include: qualitative research methods, gender, social movements, peace studies, and the sociology of emotions. She has been engaged in research on the Israeli women's peace mobilizations since the late 1980s and has published extensively and presented many papers on these topics. Her co-authored book *Iraqi Jews in Israel* won a prestigious academic prize in Israel. Currently she is a member of the board of RC 48 and a member of RC 36 and RC 06 of the ISA.

David Boyns is an associate professor of sociology at California State University at Northridge. He studies social theory, cultural sociology, media studies, and the sociology of emotions. He has published on topics like the emotional dynamics of the self, social construction processes in virtual worlds, and the sociology of deviance. His current research investigates the human–technology interface, the sociology of everyday life, and the sociology of wellness and creativity.

Natàlia Cantó-Milà is an associate professor at the Open University of Catalonia (UOC) in Barcelona. After her PhD thesis at the University of Bielefeld (Germany) on Simmel's relational sociology, she went to Leipzig, where she taught sociology for 4 years. After 10 years in Germany she returned to Barcelona and is now working in the UOC's Arts and Humanities Department. With her research group (GRECS) she holds a research project funded by the Spanish Ministry of Science and Innovation, which aims at analyzing the meaning of love and commitment in late modernity. Her last publications have been on friendship in Pro-Ana communities, on gratitude as a form of sociation, on love relationships and the uses of ICTs, and on Simmel's philosophy of money. Her main research interests are emotions as objects of sociological analysis, social theory, sociology of experience (Erlebenssoziologie), sociology of the future and future studies, the body in the social sciences, and social theory.

226 Contributors

Rebecca Chiyoko King-O'Riain is a senior lecturer at the National University of Ireland, Maynooth. Her research interests are in emotions, technology, and globalization; race/ethnicity and critical race theory; people of mixed descent, beauty, and Japanese Americans. She has published in *Ethnicities, Sociology Compass, Journal of Asian American Studies*, and *Amerasia Journal*. Her book *Pure Beauty: Judging Race in Japanese American Beauty Pageants* (2006, University of Minnesota Press) examines the use of blood quantum rules in Japanese American beauty pageants. She is currently researching and writing about 'global mixed race' and 'the globalization of love.'

Eran Fisher is assistant professor in the Department of Sociology, Political Science, and Communication at the Open University of Israel. He writes on economic, cultural, and social facets of digital technology. His book *Media and New Capitalism in the Digital Age* (2010, Palgrave), which received an Honorable Mention of the Association of Internet Research Book Award, is coming out in paperback in 2013.

Henrik Fürst is a doctoral student in sociology at Uppsala University. He is currently involved in research about publishing and digital media. He holds an undergraduate degree in sociology and a master's degree in education from Stockholm University. Previously he has been a research assistant in sociology at the Swedish National Defence College. One of his most recent publications, *Core Values and the Expeditionary Mindset: Armed Forces in Metamorphosis* (2011, Nomos), is edited with G. Kümmel, and Fürst is also a contributor.

Nina R. Jakoby, PhD, is a senior research and teaching associate at the Institute of Sociology, University of Zurich. Her main research areas are the sociology of emotions with a focus on grief and sadness, sociological theories, empirical social research, and family sociology.

Arvid Kappas is professor of psychology at Jacobs University Bremen. His research focuses on emotions for more than 25 years. Currently, he is particularly interested in interdisciplinary approaches to Internet communication, affective computing, and technology-enhanced learning. In 2011 he edited *Face-to-Face Communication on the Internet* with Nicole Krämer.

Dennis Küster is a postdoctoral fellow at Jacobs University Bremen. He obtained his PhD on the relationship between emotional experience and emotional expression. Currently, he focuses on the emotions, psychophysiology, and social context elicited by online communication. He is particularly interested in how textual statements of emotions on the Internet relate to online self-presentation, anonymity, and bodily emotional responses.

Daniele Loprieno is an instructor of sociology at College of the Canyons in Santa Clarita, California. Her research focuses on the sociology of mass media, film, Internet, and fan studies. Other research topics include sociology of health, the life course, race, and the family.

Alessandra Micallizzi obtained a PhD in Communication and New Technologies at IULM University with a final project on mourning and the practices of sharing emotions on-line. She had a post doctoral fellow in the same university until 2012. She recently collaborates with the department of Communication, Behavior and Consumption participating at academic life. She is also researcher for Episteme, a Private Institute of research. Her area of interest includes emotion and new media and above all on-line shaming and the construction of public reputation. Her last publication edited with Manuela Farinosi is titled *"Netquake: digital media and natural distasters"* and is about the sharing of emotions around the traumatic experience of the earthquake that devastated Abruzzi.

Francesc Núñez is an associate professor at the Open University of Catalonia (UOC) in Barcelona, and head of the Department of Humanities. He holds a PhD in sociology (UAB) and has a BA in philosophy and sociology. He is member of the Research Institute in Sociology of Religion (ISOR/UAB), and member of the research group 'GRECS' (Studies in Culture and Society). Together with GRECS he works on a Ministry of Science and Research-funded research project (I + D), analyzing love and commitment in late modernity. He is director of the digital revue *Digithum: The Humanities in the Digital Age*, and has published the monograph *Les Plegades: Secularized Priests* (Mediterrània Press), and several papers on online sociability.

Mervi Pantti is associate professor and director of the International Master's Programme in Media and Global Communication in the Department of Social Sciences at the University of Helsinki, Finland. She has published on mediated emotions, crisis reporting, digital visual culture, and participatory media in several international journals. Her latest books are *Amateur Images and Global News* (with Kari Andén-Papadopoulos, 2011, Intellect) and *Disasters and the Media* (with Karin Wahl-Jorgensen and Simon Cottle, 2012, Peter Lang).

Tamara Peyton (tspeyton@psu.edu) is at The Pennsylvania State University, in the College of Information Sciences & Technology. Her current work is on 'mHealth,' examining the positive social impacts of mobile technologies and gaming on adolescents with chronic health conditions. With a background in sociology, she has also considered the ways social media and digital games encourage specific forms of teamwork, leadership, and expertise.

Simone Reiser, BA in sociology, is a graduate student of sociology at the University of Zurich. Her research interests include mixed methods and emotions with particular reference to feeling rules.

Elisabetta Risi has a PhD in information society. She is assistant professor at IULM University in Milan and post-doc researcher at Communication Institute in the same university. Recent research interests include sociology of emotion, social media studies, and works on movement emotions and feelings rules. Correspondence concerning this chapter should be addressed to Elisabetta Risi, Istituto di Comunicazione, Università IULM Milan, Via Carlo Bo, 1 -20143. Email: elisabetta.risi@gmail.com

Swen Seebach is a PhD candidate and researcher at the Internet Interdisciplinary Institute (IN3) in Barcelona. He is writing his PhD thesis on the subject of love relationships, love rituals, consumption, and electronic communication. Together with his research group (GRECS) he works on a Ministry of Science and Research-funded research project (I + D), analyzing love and commitment in late modernity. His central research interests are emotions as objects of sociological analysis, the body in times of the Internet, emotions and the body, and time and the future as sociological research objects. In his last publication, *Ana's friends. Friendship in Online Pro-Ana Communities*, he and his colleague Natàlia Cantó Milà have analyzed Pro-Ana forums on the Internet.

Frances Shaw is a research assistant at the University of Sydney. She recently completed her PhD in politics and international relations at the University of New South Wales, Sydney, with a thesis on feminist online networks in Australia. She is currently working on a research project on Australian Internet histories with Professor Gerard Goggin, while also developing new research in memetic politics.

Andrea L. Stanton is assistant professor of Islamic Studies at the University of Denver. Trained as a historian, her work focuses on Islam in the Middle East and elsewhere in the 20th–21st centuries. Her research examines expressions of faith and religious identity in print and broadcast media, and investigates the sometimes conflictual, sometimes cooperative relationship between new technologies and claims to religious authority. She serves on the board of the Syrian Studies Association, as editor of its bi-annual *Bulletin*, and as editor of H-Levant, a scholarly list server with over 1,100 members.

Jakob Svensson is holding a position as assistant professor in the Department of Media and Communication Studies at Karlstad University, Sweden (www.kau.se/en/media). He is also director of the research network

Contributors 229

HumanIT (www.kau.se/en/humanit). He obtained his PhD in 2008 from Lund University, Sweden, with a dissertation on citizenship practices and civic identities. His research continues to focus on political communication from participatory and civic perspectives. His current research interests include online socio-political practices and online identity formations, and their dialectical intertwining with offline territories. Svensson is member of various international networks and conference program committees that explore communication and new media aspects of democracy and participation, and has been called in by the Swedish government to give his views on e-democracy projects and proposals.

Minttu Tikka is a doctoral student in the department of Social Research/ Media and Communication Studies at the University of Helsinki, Finland. Her research interests include new media and crisis communication.

Author Index

A
Anderson, J. 80, 83, 94–95, 97

B
Bakardjieva, M. 196, 208
Barabasi, A-L. 50, 58
Barbalet J. M. 2, 13, 164–165, 169, 172, 175
Bauman, Z. 4, 13, 109, 111, 158, 163, 176
Beck, U. 20, 29–30, 162–165, 176, 178, 190
Benski, T. 1, 164, 169, 176, 221
Boccia, A. G. 196, 208
Boltanski, L. 181, 185, 190
Bolter J. D. 208
Bourdieu, P. 5, 13, 22, 26–27, 30
boyd, d. 19, 21, 24, 30–31, 167, 176, 195, 208
Bruner, J. 161, 166, 168, 176, 208
Bunt, G. 80, 84, 94–95, 97

C
Castells, M. 4, 13, 18, 21, 30, 172, 176
Chmiel, A. 50–51, 56, 59
Chouliaraki, L. 179–182, 189, 190–191
Collins, R. 2–3, 7, 13, 22, 26–27, 30, 33–35, 37, 41–43, 45–46, 164, 172, 176

D
Dayan, D. 206, 209
Della Porta, D. 172, 175–176
Derrida, J. 114–115, 127
Ditton, T. 33, 39, 46
Dodge K. A. 208–209
Durkheim, E. 2, 13, 34, 45, 158, 162

E
Eickelman, D. 83, 95, 97
Elias, N. 6, 13, 18–19, 22–25, 27–30

F
Flichy, P. 195, 209
Foucault, M. 18, 23–24, 31

G
Galloway, A. R. 116, 127
Golder, S.A. 49, 54, 59
Grusin R. 208
Gump B. B 194, 209

H
Habermas, J. 81, 93
Haidt J. 208, 210
Haviland-Jones J.M. 208–210
Hochschild, A. R. 2–3, 6, 13, 17, 21–22, 24–25, 27, 31, 67–68, 75, 78, 102, 107, 112, 131, 142, 169, 171, 176
Horton, D. 33, 36, 46
Huizinga, J. 41, 46

I
Illouz, E. 33, 99, 103, 107, 112, 146, 158
Ismail, S. 93–94, 97

K
Kappas, A. 7, 8, 48, 55–61, 226
Katz E. 206, 209
Kazys, V. 196, 209
Kemper, T. D. 2, 13, 14, 22, 176, 194, 209
Krumhuber, E. G. 53, 59
Kulik J. H. 194, 209

L
Lemerise E.A. 208–209

232 Author Index

Lewis M. 204, 208–210
Lewis M. 208–210
Lombard, M. 33, 39, 46
Luminet, O. 194, 209

M
Mahmood, S. 81, 93–94, 97–98
Mauss, I. B. 53, 55–56, 59
McCauley C. 208, 210
Mead, G. H. 66, 78, 102, 112
Micalizzi, A. 197, 202, 209
Mol, A. 124–126, 128

O
Oham A. 208, 209
Ong, W. J. 127–128, 195, 209

P
Palczewski, C. 212, 224
Pandolfi M. 204, 207, 209
Parrott, W. G. 204, 209
Pata K. 196, 198, 209
Pennebacker J. W. 49, 194–196, 202, 209
Peters, J.D. 195, 209

R
Rancière, J. 222
Reisenzein, R. 53, 57, 60

Rimè B. 162, 166–168, 177, 193–196, 206, 208, 209
Riva, G. 37, 40, 47, 207, 209
Rowbottom, J. 205, 210
Rozin P., Haidt J. 208, 210

S
Scheff, T. J. 164, 169, 173, 177
Shott, S. 194, 210
Simmel, G. 108, 112, 145–146, 158, 225
Skitka, L. J. 49, 58, 60
Spackman M. 204, 209
Strauss, A. L. 100–102, 111–112
Suchman, L. 124–125, 128

T
Thelwall, M. 49, 51, 56, 59–61
Thoits, P. 114, 126, 128
Thompson, J. B. 208, 210
Tidwell, L.C. 49, 53, 61
Tomlinson, B. 212–213, 216, 222, 224
Turkle, S. 34, 36–37, 47, 196, 207, 210
Turner, V. 146, 158

W
Walzer J. 195, 210
Wohl, R. 33, 36, 46

Subject Index

A
Adaptive function 194
Age of the relationship 149–150
Appeal video 11, 178–180, 182–183, 186–187, 189–190
Arabic 86–93, 96
Arguments 13, 93, 148, 155, 213
Attention 10, 34, 35, 38, 50, 67, 83–85, 121, 125, 138, 145, 151, 168, 186, 190, 1193–195, 197, 202–203, 206, 214–215, 219–220
Australia 5, 140–141, 211
  Australian 80, 83, 94–95, 98, 132, 211–213, 218–219
Automated text analyses 49–55
Aversive politics 212

B
Backchannels 217–218, 220–221
Black box 116–117
Blog 3–5, 11–13, 30, 49–50, 92, 96, 132, 134, 143, 161, 167–169, 175, 182, 211–224
  Bloggers 12–13, 21, 59, 132, 142, 211–222
  Feminist 5, 12, 212, 215–217, 224
Budget 147
Broadband Visual Communications 134–135, 140–141, 143

C
Civility 5, 213
Cognitive rumination 194–196, 206
Cohabitation 148–151, 153–154
Commercialization/Commercialized/ Commercializing 5, 13, 31, 78, 99, 107, 111–112, 142, 176, 184
Commitment 3, 5, 9–10, 65, 93, 99–100, 102–103, 108–109, 126, 144, 146, 214, 220, 223
Communication 1–6, 10–12, 18–20, 30–32, 35–39, 43, 45–51, 53–56, 58–59, 61, 68–69, 74, 77, 80–85, 93–98, 107, 112, 128, 131–132, 134, 138, 141–158
  Failure of 149
  Phatic 20, 31, 133, 143
Community 5, 12, 13, 42–43, 46–47, 50, 56–57, 59, 68, 80–81, 85, 91, 93–94, 96, 98, 120–121, 132, 171, 179–180, 188, 211-212, 214–223
Compassion 178–179, 187–188, 191, 202
Constructing memory 204
Convergence culture 14, 36–37, 44, 46
Co-presence 2–3, 7, 33–38, 40–42, 46
Cosmopolitan empathy 11, 178, 182, 189
Cosmopolitanism/Cosmopolitan 11–12, 178–179, 182, 185, 188–191
Counterhegemonic politics 211
Cyber-practice 193
(Cyber) Voyeurism 12, 193, 196–198, 206–207

D
Daily tasks 147, 149–150, 156–157
Deep acting 22, 68, 171
Desires 142, 145–146, 149, 150, 152–154, 156–157, 164–165
Différance 114, 118
Digital literacy 25–27, 31
Digital memories 198

234 Subject Index

Digitality 117, 126, 128
Disability 218
Disaster communication 11, 178–180, 189
Disciplining 23–24, 32
Discursive politics 215, 222
Dissensus 222, 224
Distance (Distant/Distances) 10, 29, 35–36, 46, 55, 68, 73, 131–134, 136, 138–139, 141–142, 144–145, 147–148, 150–151, 153–154, 156–158, 163, 167, 170–171, 173, 177, 180–182, 188, 190, 192

E
Email 3,4, 39, 40, 41, 49, 81, 84–85, 99–100, 131–132, 137, 144, 147–148, 151–157, 172, 198, 206, 217
Emoticon 7–9, 21, 30, 53–54, 80–98, 152
  Smiley 54, 80–82, 84–85, 87, 89–92, 96
Emotions
  Anger 12–13, 149, 152, 164, 165, 168, 169, 171, 179, 193, 201, 202, 204–205, 209, 212–214, 221, 222
  Disgust 12, 201, 204–205, 208, 210, 214
Emotional
  career 103, 108–109
  communication 49, 151–152
  display 8, 102
  energy 7, 22–23, 26–27, 29, 34–35, 41, 43–45, 170, 172, 176
  labor 21, 102, 107
  mobilization 178
  ordering 9, 102–103, 107, 111
  public sphere 179, 189, 191
  regime 178, 180–182, 189
  socialization 9, 67, 99, 103, 108–111
Emotions sharing 168–169, 171–172
Emotive-mediated quasi- interactions 208
Empathic role-taking sentiments 194
Empathy 11–12, 57, 151, 178–179, 181–183, 185, 187, 189
  Expression of 76, 152, 154, 183, 194
  Fear 17, 68, 76, 102, 203, 205–206, 208–209
  Frustration 10, 12–13, 140, 149, 152, 166, 168, 174, 213, 215

Grief 4, 8, 65–79, 201–202, 205
Happiness 5, 9, 54, 75, 91, 99, 105–107, 109–110, 157, 177
Hate 12, 141, 168, 201, 204
Hope 5, 9, 17, 23, 44, 72–73, 99, 101, 103, 105–110, 144–145, 157, 184, 199, 203, 205, 214
Joy 75, 152, 157
Joyful 185, 188
Mood 28, 34–35, 38, 49–50, 54, 59, 137, 162, 165–166, 168–169, 204
Online emotions 45, 48–51, 53–55, 57, 158
Reflexive role-taking emotions 194
Resentment 11, 161–177
Ressentiment 163, 176–177
Sentiment 28, 51, 56, 60–61, 114–115, 124, 127, 161, 179, 193–194, 204–205, 207–208
Shame 12, 54, 164, 168, 173, 181, 187, 193–194, 201, 204–205, 207, 209
Suffering 12, 68, 75, 133, 142, 166, 169, 178–190, 192, 195–196, 201–202, 205
Unhappiness 84
Evaluation 5, 27, 32, 55, 61, 104, 109, 175
Everyday life practices 146
Expectations 17, 65, 68–69, 108, 145–146, 149, 152–154, 156–157, 163, 172, 184
Expressive rationality 23
Extraordinary moments 146

F
Facebook 1–3, 6, 9–10, 19, 32, 49–50, 56, 60, 76, 93, 113–114, 117–119, 123–127, 131–133, 142, 144, 167–168, 172, 197, 199, 220
Face-to-face (F2F) interaction 2, 7, 33, 36, 40–44, 72, 81, 132
  non- 35, 39–41
Face-to-face communication 58–59, 151, 182
Face-to-face 2, 4, 7, 45, 59, 72, 81, 132, 136–137, 145, 148, 154–155, 171, 195
Flaming 199
Forgiveness 86–87, 205
Forum 4–6, 11, 40–42, 49–50, 56, 59–61, 68, 73–75, 77, 80–81,

83, 85–91, 93–98, 113, 117, 127, 167–169, 217
Fragment-de-vie 197, 207
Fuck Politeness 212–213, 223

G
Gender 6, 31, 45–46, 53, 61, 89, 91–95, 98, 121, 127, 136
Globalization 132, 143, 179
Gold standard 52, 55–56

H
Habitus 26, 27, 135
Heart rate 56–57
Hijab 89
Hoyden About Town 215, 217, 219
Humanitarian 11, 12, 178–192
Hybrid narrative environment 196, 198

I
Immediacy 11, 151
Immersion 40, 42, 44, 207, 209
  Emotional immersion (subset) 38
Individualization 18, 20–21, 29, 65, 76, 79, 162, 165
Instant messages 148, 151
Intensification 216
Interaction Ritual Theory (IRT) 33–37, 41, 43–44, 46
Interaction Rituals (IRs) 3, 7, 13, 22, 33–35, 37–45
Technologically mediated interaction (TMI) 7, 34–36, 38, 39, 40, 41, 42, 44
Internet dating 9, 99–112, 146
(Internet) trolls 12, 212, 215, 217, 223
(Internet) trolling 215–217, 220
Ireland 131, 134–143
Islam 80, 82–83, 86–87, 90–92, 94–98
  Islamic Emoticon 8–9, 80–81, 87–94, 96
  Islamic 80–81, 83–84, 86–98
  Islamism 94, 97
  Muslim 8–9, 80–88, 90–91, 93–94, 96–98

J
Job Insecurity 4, 11, 161,162–163, 167, 172, 175

K
Knowledge genres 125–126

L
League of Legends 120,122, 128
Like button 9, 10, 113, 115–119, 125, 126
Liminality 115, 146, 154, 158
Linguistic Inquiry and Word Count (LIWC) 49, 51, 53, 60
Long distance relationships 145, 150–151, 153, 156–157

M
Magic Circle 41–42
Massively Multiplayer Online Role Playing Games (MMOs) 40, 43
Mimesis 195
Mixed 3, 59, 91–93, 131, 134–136, 139, 141
Mobile phone 5, 20, 31, 144–145, 147–151, 154
Moderation 5, 12, 28–29, 212, 215, 217, 223
Monologist dialogue 202
Mother 131, 140, 188, 197, 214
MySpace 19, 56, 60, 61

N
Naming and shaming 205, 210
Narrative
  Analysis 168, 177, 198
  nature 196
  network analysis 197, 206, 209
Neoliberal/Neoliberalism 114, 116–118, 123–128
Network society 4, 5, 13, 18, 21, 24, 27, 29–30, 32, 176

O
Online harassment 215–216
Online rituals 154
Ontological politics 113, 116, 124, 126, 128
Outlet 211–215, 222

P
Panopticon 23
Parasocial Interaction 33, 36–38, 40, 42–44, 46
  Parasocial Presence 7, 33, 37–38, 40, 43
Parasociality 33, 36, 39, 43–44
Phantasmagoria 5, 100, 102, 105–109
Physiological measures 54, 56–57
Physiological responses 48, 59
Piety 9, 80–81, 83–84, 93–95, 97–98

Pinterest 117
Politeness 5, 54, 57
  Dissonance 215, 221
Political subjectivity/subject 13, 215, 222
post-Humanitarian 11, 181–182, 189, 191
Power 5–7, 13, 17–18, 23–25, 27–32, 37, 40, 50, 60, 83, 92, 101, 115–116, 123–124, 127–128, 142–143, 165, 169, 173, 178, 194, 196, 202, 213, 218–219, 222
Precarious 4, 11, 162–163, 167–174
Presence 2, 3, 7, 10, 33–47, 68, 73, 95, 98, 117, 134, 145, 166, 173, 180, 184, 194–196, 207
  Social Presence 7, 37–40, 43, 45–47
Private media events 206
Process of civilization 24, 27–29
Prosumer 103, 107, 112
Prudery 195–196, 200
Public keyhole syndrome 207
  Wechselwirkung 145, 156

R

Recognition 6–7, 22–24, 26–27, 85, 162, 172–173
Religion/Religious 9, 13, 45, 72, 75, 81–84, 88, 90, 93–95, 97, 98,136, 158, 165
Remembrance 65, 69, 74, 146, 202, 204–205
Respect 35, 68, 90, 93, 100–101, 204–206, 223
Revenge 201, 204–207
Rupture 222

S

Safe space 5, 13, 211–212, 215–216
Salafi 91
Secondary orality 195
Seduction 152–154
Self-regulation 152, 176
SentiStrength 49, 51, 53, 208
Sex 42–44, 67, 94–95, 98–99, 104, 111, 120–121, 135–136, 150, 152, 155
Simulation/Simulate 7, 33, 36–39, 44, 150, 155
Situated action 116–117, 124–125, 128
Skype 1–5, 10, 131, 133–142, 147–148, 150–152, 155, 157

Social
  media 6, 11, 19, 32, 58–59, 60, 83, 113, 116–118, 161, 168, 183
  Sharing 12, 113, 161–162, 166–167, 171–172, 174, 177, 193–195, 206–207, 209
  world/arena 33–34, 42, 44–45, 99–104, 107, 110–112
  arena 5, 23, 27, 42, 100, 101
  sharing of emotions 12, 162, 166–167, 171–172, 174, 193–195, 206
Socio-cultural artifact 195
Sociotechnical 113, 115–116, 126
Solidarity 2–3, 34–35, 41, 46, 117, 120, 127, 170, 178, 184–186, 188, 217, 221
Specialized uses of electronic communication (devices) 147, 157
Subjective measures 56

T

Taken-for-grantedness 145, 157
Temporality 8, 122, 156
  Future 5, 9, 13, 25, 29, 35, 44, 57–58, 66, 74, 76, 101–110, 142, 153–154, 162, 170, 190, 224
  Present 105, 107–108, 110, 153, 161, 164, 165, 175, 189, 195, 197, 205, 208, 219
Transconnective Space 3, 10, 131, 136
Transconnectivity 135, 139
Traumatic events 194, 196
Trigger warnings 12 ,222
Tumblr 218, 221
Twitter 6, 19, 49–50, 59–60, 117, 133, 167, 217–218, 220–221

U

Ulama 83, 90
Uncertainty 104, 106, 110, 118
Usenet 217
User generated content 11, 60, 161, 175, 178, 180, 182, 187–189

V

Visual contact 151
Volunteer/ism 178, 180, 191

W

Witnessing 18, 37, 179, 182, 187–188, 191, 209
Written self-disclosure 196
Written word 144, 148